Wetlands as Nature-Based Solutions: Protecting Wetland Ecosystems for Climate Resilience, Water Management, Biodiversity Conservation, and Sustainable Development Worldwide

I0029960

Copyright

Wetlands as Nature-Based Solutions: Protecting Wetland Ecosystems for Climate Resilience, Water Management, Biodiversity Conservation, and Sustainable Development Worldwide

© 2025 Robert C. Brears

The author and publisher are of the same opinion regarding the views and content expressed in this work.

Disclaimer: The information in this book is provided for general knowledge and educational purposes only. While every effort has been made to ensure accuracy, the author and publisher make no representations or warranties with respect to the completeness or suitability of the content. The author and publisher accept no liability for any errors, omissions, or outcomes resulting from the application of information contained herein. Readers are advised to consult appropriate professionals or authorities before acting on any material presented.

ISBN (eBook): 978-1-991369-81-9

ISBN (Paperback): 978-1-991369-82-6

Published by Global Climate Solutions

First Edition, 2025

Cover design and interior layout by Global Climate Solutions

Table of Contents

Introduction

Nature-Based Solutions (NbS) have emerged as essential tools for addressing global environmental challenges, offering sustainable approaches that harness natural processes to enhance ecosystem resilience. Among these solutions, wetlands play a crucial role in providing a wide range of ecosystem services, contributing to climate resilience, biodiversity conservation, and sustainable development. Wetlands are among the most productive ecosystems on Earth, supporting diverse flora and fauna while regulating hydrological cycles and storing carbon. Their ability to mitigate flooding, filter pollutants, and sustain livelihoods makes them vital assets in environmental management.

Despite their ecological and societal benefits, wetlands face significant threats from human activities, including land conversion, pollution, and climate change. Urban expansion, agricultural intensification, and infrastructure development have led to large-scale wetland degradation, reducing their capacity to deliver critical ecosystem functions. Climate change further exacerbates these pressures, altering precipitation patterns, increasing temperatures, and intensifying extreme weather events that impact wetland hydrology. The loss of wetlands not only diminishes biodiversity but also weakens natural resilience against environmental disruptions. As global commitments to sustainability and climate adaptation increase, integrating wetlands into policy and planning as a NbS becomes essential.

This book explores the role of wetlands as a nature-based solution by examining their ecological significance, their contributions to climate resilience, and their integration into sustainable management strategies. It provides an in-depth analysis of wetland ecosystems, detailing their functions in carbon sequestration, flood mitigation, and water purification. The book also discusses their importance for biodiversity, highlighting the interdependence between wetland habitats and species conservation.

Beyond their ecological benefits, wetlands offer economic and social advantages. Their capacity to improve water quality reduces the need for costly engineered solutions, while their role in supporting fisheries, agriculture, and ecotourism provides economic opportunities for communities. The economic valuation of wetlands is increasingly recognized in environmental finance, guiding investment decisions in conservation and restoration projects. Governments and international organizations have begun integrating wetland conservation into climate policies, recognizing the cost-effectiveness of NbS compared to traditional infrastructure.

However, effective wetland conservation and restoration require robust governance, stakeholder engagement, and policy alignment. The book examines the challenges of wetland management, from land-use conflicts to regulatory frameworks, and explores how international agreements such as the Ramsar Convention shape global conservation efforts. Additionally, it highlights the importance of local participation, emphasizing the role of communities, indigenous knowledge, and cross-sector collaboration in sustainable wetland management.

By providing a comprehensive overview of wetlands as a NbS, this book aims to inform policymakers, researchers, and practitioners on strategies for enhancing wetland conservation and restoration. It advocates for integrating wetland-based approaches into climate adaptation and development planning, ensuring long-term environmental and societal benefits. As global efforts to mitigate climate change and biodiversity loss accelerate, recognizing the value of wetlands will be critical in building resilient and sustainable ecosystems.

Chapter 1: Wetland Ecosystems: An Overview

Wetlands are vital ecosystems that support biodiversity, regulate water cycles, and provide essential ecosystem services. Found in coastal and inland regions, they store carbon, filter pollutants, and mitigate climate impacts. Despite their value, wetlands face threats from land conversion, pollution, and climate change, leading to significant losses worldwide.

This chapter explores the types of wetlands, their ecological characteristics, and their global distribution. It highlights their critical functions and the challenges they face, emphasizing the need for conservation and integration into sustainable environmental strategies.

Types of Wetlands

Wetlands are diverse ecosystems characterized by the presence of water, either permanently or seasonally, and support unique plant and animal life. They vary widely in size, function, and ecological characteristics, but all provide essential ecosystem services such as water filtration, flood control, carbon sequestration, and biodiversity support. Wetlands can be broadly categorized into coastal wetlands, inland wetlands, and artificial wetlands, each playing a distinct role in environmental sustainability.

Coastal Wetlands

Coastal wetlands are found in areas where land meets the ocean, including estuaries, salt marshes, and mangrove forests. These wetlands are influenced by tidal movements and are highly productive ecosystems that provide habitat for many marine and terrestrial species.

- **Mangroves**
  Mangrove forests thrive in tropical and subtropical coastal regions where saltwater and freshwater mix. They consist of salt-tolerant trees and shrubs that stabilize coastlines, reduce storm surges, and provide breeding grounds for fish, crustaceans, and birds. Their dense root systems trap sediments, improving water quality and preventing coastal erosion. Mangroves are crucial for carbon storage, making them vital for climate change mitigation.
- **Salt Marshes**
  Found in temperate and subarctic regions, salt marshes are coastal wetlands dominated by salt-tolerant grasses, sedges, and shrubs. They play a key role in buffering coastal areas against storm surges and sea-level rise by absorbing excess water. Salt marshes also act as natural water filters, trapping pollutants and improving coastal water quality. They provide essential feeding and nesting sites for migratory birds and serve as nurseries for fish and shellfish.
- **Seagrass Meadows**
  Seagrass meadows grow in shallow, coastal waters and support a diverse range of marine life, including sea turtles, dugongs, and juvenile fish. They help stabilize sediments, improve water clarity, and store large amounts of carbon in their root systems. Seagrass meadows are among the most productive and threatened ecosystems, often impacted by coastal development, pollution, and climate change.

Inland Wetlands

Inland wetlands occur away from coastal regions and include marshes, swamps, bogs, fens, floodplains, and peatlands. These wetlands provide critical ecosystem services such as groundwater recharge, carbon sequestration, and flood mitigation.

- **Marshes**
  Marshes are wetlands dominated by herbaceous plants rather than woody vegetation. They are often found along riverbanks, lakes, and low-lying areas. Marshes provide

essential habitat for amphibians, waterfowl, and fish while improving water quality by filtering sediments and nutrients. There are freshwater and saltwater marshes, both of which play a role in regulating hydrological cycles.

- **Swamps**
  Swamps are wetlands dominated by trees and shrubs, commonly found in flood-prone areas and along rivers. They support rich biodiversity, including fish, reptiles, mammals, and birds. Swamps act as natural flood buffers, absorbing and slowly releasing excess water. They also filter pollutants and contribute to carbon sequestration. Examples include cypress swamps in North America and tropical peat swamps in Southeast Asia.

- **Bogs**
  Bogs are nutrient-poor wetlands with acidic, waterlogged conditions, often formed in cool climates where peat accumulates over time. They support unique plant species such as sphagnum moss and carnivorous plants. Bogs store large amounts of carbon, making them significant for climate regulation. However, they are highly sensitive to environmental changes and human disturbances such as peat extraction and drainage.

- **Fens**
  Fens are similar to bogs but have higher nutrient levels due to groundwater influence. They support a greater diversity of plant and animal species. Fens play an important role in filtering water and maintaining hydrological balance in their surroundings. Like bogs, they are highly vulnerable to degradation from land-use changes and water extraction.

- **Floodplains**
  Floodplains are wetlands formed along rivers and streams that experience periodic flooding. They support fertile soils and diverse vegetation, making them important for agriculture and wildlife habitat. Floodplains help mitigate flood damage by absorbing excess water and reducing the risk of downstream flooding. However, they are often altered by human activities, including dam construction and urban expansion, which disrupt their natural functions.

Artificial Wetlands

Artificial or constructed wetlands are human-made systems designed to mimic the functions of natural wetlands. They are increasingly used for wastewater treatment, stormwater management, and habitat restoration.

- **Constructed Treatment Wetlands**
  These wetlands are engineered to improve water quality by removing pollutants, excess nutrients, and sediments. They use wetland vegetation and microorganisms to break down contaminants, providing a cost-effective and sustainable alternative to conventional wastewater treatment plants. Constructed wetlands are widely used in urban areas, industrial facilities, and agricultural landscapes.
- **Reservoir Wetlands**
  Some reservoirs and man-made lakes develop wetland characteristics over time, supporting aquatic vegetation and wildlife. These wetlands help regulate water storage, reduce sedimentation, and enhance biodiversity. However, they require careful management to balance water resource demands and ecological functions.
- **Restored Wetlands**
  Restoration efforts aim to recreate or rehabilitate degraded wetlands to recover their ecological functions. Restored wetlands provide habitat for wildlife, improve water quality, and enhance flood resilience. Governments and environmental organizations increasingly recognize the importance of wetland restoration as part of broader conservation strategies.

Ecological Characteristics

Wetlands are unique ecosystems defined by their hydrology, soil composition, and vegetation. Their ability to support diverse plant and animal life makes them one of the most biologically productive ecosystems on Earth. The ecological characteristics of wetlands influence their functions, including water filtration, carbon sequestration, and flood control. Understanding these characteristics is essential for effective conservation and sustainable management.

Hydrology

Hydrology is the defining feature of wetlands, distinguishing them from other terrestrial and aquatic ecosystems. Wetlands exist where water saturates the soil either permanently or seasonally, creating conditions that support specialized plant and animal life.

- **Water Sources**
  Wetlands receive water from precipitation, groundwater, surface runoff, and tidal movements. The water balance in a wetland is influenced by seasonal fluctuations, climate, and topography. Some wetlands, such as bogs, primarily rely on rainfall, while others, like floodplains, depend on river overflow.
- **Hydroperiod**
  The hydroperiod refers to the seasonal pattern of water presence in a wetland. It determines the wetland's biological composition and ecological functions. Some wetlands, such as marshes, remain flooded year-round, while others experience periodic drying, as seen in vernal pools and floodplains. The duration and frequency of flooding shape the wetland's ability to support different species.
- **Water Flow and Connectivity**
  Wetlands can be isolated or connected to larger water systems, such as rivers, lakes, or coastal areas. This connectivity influences nutrient cycling, species migration, and the wetland's role in regulating hydrological processes. Wetlands that are linked to river systems contribute to groundwater recharge and flood mitigation, while isolated wetlands provide localized water retention.

Soil Composition

Wetland soils, also known as hydric soils, are characterized by prolonged water saturation, leading to low oxygen levels. This creates unique chemical and biological conditions that influence plant growth and nutrient cycling.

- **Organic and Mineral Soils**
  Wetland soils can be classified into organic and mineral soils based on their composition. Organic soils, found in peatlands and bogs, contain high amounts of decomposed plant material, making them significant carbon sinks. Mineral soils, present in marshes and floodplains, have a mix of sand, silt, and clay, with varying levels of nutrients and water-holding capacity.
- **Anaerobic Conditions**
  Due to water saturation, wetland soils often experience anaerobic (low oxygen) conditions. This affects microbial activity and nutrient availability. In the absence of oxygen, microorganisms break down organic matter slowly, leading to the accumulation of peat in some wetlands. The anaerobic environment also influences nutrient cycling, particularly nitrogen and phosphorus dynamics.
- **Soil pH and Nutrient Availability**
  Wetland soil pH ranges from highly acidic in bogs to more neutral or alkaline in fens and marshes. Acidic conditions in bogs limit microbial decomposition, resulting in peat accumulation. Fens and marshes, enriched by groundwater and surface water, tend to have higher nutrient levels, supporting diverse plant and animal communities.

Vegetation

Wetland vegetation is adapted to fluctuating water levels and low-oxygen conditions. Plants in wetlands provide habitat structure, stabilize soil, and contribute to nutrient cycling. Wetland plant communities are classified into emergent, floating, and submerged vegetation.

- **Emergent Vegetation**
  Emergent plants grow in shallow water or along wetland margins, with their roots submerged and stems rising above the water. Examples include reeds, cattails, and sedges. These plants help stabilize soil, filter pollutants, and provide food and shelter for wildlife.
- **Floating Vegetation**
  Floating plants either float freely on the water surface or have roots anchored in the substrate. Species such as water lilies, duckweed, and water hyacinth are common in wetlands. They provide shade, reduce water temperature fluctuations, and support aquatic food webs.
- **Submerged Vegetation**
  Submerged plants grow entirely underwater, playing a key role in oxygenating water and stabilizing sediments. Examples include eelgrass and pondweeds. These plants provide habitat for fish, invertebrates, and other aquatic species while contributing to nutrient cycling.

Biodiversity and Trophic Interactions

Wetlands support a diverse array of species, from microorganisms to large mammals. The complex food web in wetlands includes primary producers, herbivores, carnivores, and decomposers.

- **Primary Producers**
  Wetland plants and algae serve as primary producers, converting sunlight into energy through photosynthesis. They provide food for herbivores and contribute organic matter to the ecosystem.
- **Herbivores and Detritivores**
  Herbivores, such as insects, fish, and mammals, consume wetland vegetation. Detritivores, including insects and crustaceans, break down organic matter, recycling nutrients within the ecosystem.
- **Carnivores and Apex Predators**
  Wetlands support predators such as amphibians, birds, and

mammals. Species like herons, otters, and alligators play essential roles in maintaining ecological balance.

Global Distribution and Threats

Wetlands are found on every continent except Antarctica, covering approximately six percent of the Earth's land surface. Their distribution varies based on climate, geography, and hydrological conditions. Some wetlands, such as tropical mangroves and peatlands, thrive in warm and humid regions, while others, like boreal bogs and high-altitude fens, exist in colder environments. Wetlands provide essential ecosystem services, but they are among the most threatened ecosystems globally. Human activities, climate change, and land-use changes have led to significant wetland degradation, resulting in biodiversity loss and diminished ecological functions. Understanding the distribution of wetlands and the threats they face is essential for developing effective conservation and management strategies.

Global Distribution of Wetlands

Wetlands occur in diverse landscapes, ranging from coastal zones to inland freshwater systems. Their distribution is influenced by regional climate patterns, hydrology, and topography.

- **North America**
  North America has extensive wetland systems, including the Everglades in Florida, Canada's boreal peatlands, and the Mississippi River floodplains. These wetlands support migratory birds, freshwater fisheries, and carbon storage. Many wetlands in the region have been lost due to agriculture, urban expansion, and industrial development. Conservation efforts, such as the protection of Ramsar sites and wetland restoration programs, aim to reverse some of this degradation.
- **South America**
  The Amazon River basin and the Pantanal in Brazil, Bolivia,

and Paraguay are among the largest wetland systems in the world. The Pantanal, the world's largest tropical wetland, supports an incredible diversity of wildlife, including jaguars, capybaras, and caimans. South America's wetlands play a vital role in hydrological cycles and carbon sequestration. However, deforestation, cattle ranching, and infrastructure development pose significant threats to their survival.

- **Europe**
  Europe's wetlands include coastal marshes, peatlands, and floodplains. Notable examples include the Danube Delta, the Camargue wetlands in France, and the peat bogs of Ireland and Scotland. Many European wetlands have been drained for agriculture and urban expansion, but restoration projects aim to improve water quality, restore biodiversity, and enhance climate resilience. The European Union's Natura 2000 network provides protection for many wetland sites.

- **Africa**
  Africa's wetlands include the Okavango Delta in Botswana, the Sudd wetlands in South Sudan, and Lake Victoria's surrounding marshes. These ecosystems support diverse wildlife, including elephants, hippos, and numerous bird species. Wetlands in Africa are crucial for water supply, food security, and climate resilience. However, they face pressures from agriculture, hydroelectric projects, and population growth. Conservation initiatives, such as community-based management programs, are working to safeguard these ecosystems.

- **Asia**
  Asia has a wide variety of wetlands, including the Sundarbans mangroves in Bangladesh and India, the Mekong Delta in Vietnam, and China's Poyang Lake wetlands. These wetlands support fisheries, rice agriculture, and biodiversity. However, rapid industrialization, land reclamation, and pollution are leading to significant wetland losses. The Ramsar Convention and regional conservation efforts aim to protect key wetland sites.

- **Australia and the Pacific**
  Australia is home to unique wetlands, such as the Kakadu

wetlands in the Northern Territory and the Coorong wetlands in South Australia. These ecosystems are vital for indigenous communities, water regulation, and biodiversity conservation. Wetlands in Pacific island nations, such as mangroves and coral reef-associated wetlands, provide coastal protection against rising sea levels. Climate change, invasive species, and water extraction are key threats to wetlands in this region.

Major Threats to Wetlands

Despite their ecological importance, wetlands are among the most threatened ecosystems worldwide. Human activities and environmental changes have led to widespread degradation, resulting in habitat loss, declining water quality, and reduced ecosystem functions.

- **Land Conversion and Urbanization**
  One of the most significant threats to wetlands is land conversion for agriculture, urban expansion, and infrastructure development. Wetlands are often drained to create farmland, industrial zones, and residential areas. This alters hydrological cycles, reduces biodiversity, and disrupts ecosystem services. In many cases, once wetlands are lost, they are difficult to restore to their original state.
- **Pollution and Water Contamination**
  Wetlands act as natural water filters, but excessive pollution can overwhelm their ability to process contaminants. Agricultural runoff, industrial discharge, and untreated sewage introduce nutrients, heavy metals, and toxic substances into wetland systems. Excessive nutrients from fertilizers cause eutrophication, leading to algal blooms that deplete oxygen levels and harm aquatic life. Heavy metals and persistent organic pollutants accumulate in sediments, affecting both wildlife and human health.
- **Climate Change**
  Climate change is altering wetland ecosystems through rising temperatures, shifting precipitation patterns, and extreme

weather events. Sea-level rise threatens coastal wetlands, leading to saltwater intrusion and habitat loss. In arid regions, prolonged droughts reduce wetland water levels, affecting biodiversity and ecosystem services. Changes in temperature and rainfall patterns can also disrupt the life cycles of wetland-dependent species.

- **Water Diversion and Hydrological Changes**
  Dams, irrigation projects, and groundwater extraction modify natural water flows, reducing the availability of water for wetlands. Many riverine and floodplain wetlands depend on seasonal flooding to maintain their ecological functions. The construction of dams disrupts these natural cycles, leading to habitat fragmentation and reduced wetland resilience. Groundwater depletion also affects wetlands that rely on subsurface water sources.

- **Invasive Species**
  Invasive plants and animals pose a serious threat to wetland ecosystems by outcompeting native species, altering nutrient cycles, and changing habitat structure. Non-native aquatic plants, such as water hyacinth and common reed, can overgrow wetland areas, reducing biodiversity and blocking waterways. Invasive fish, amphibians, and invertebrates disrupt local food webs and impact native species. Controlling invasive species requires coordinated management efforts and ongoing monitoring.

- **Unsustainable Resource Extraction**
  Overfishing, peat extraction, and excessive logging impact wetland ecosystems. Overfishing disrupts aquatic food webs, while peat extraction for fuel and agriculture contributes to carbon emissions and wetland degradation. Mangrove deforestation for timber and aquaculture reduces coastal protection and biodiversity. Sustainable resource management practices are necessary to balance economic needs with wetland conservation.

Conservation and Management Challenges

Efforts to protect wetlands are complicated by competing land-use demands, weak enforcement of environmental regulations, and a lack of awareness about wetland values. Many wetlands are located in areas of rapid development, making them vulnerable to economic pressures. Additionally, wetland restoration projects require long-term commitment and significant investment. Governments, conservation organizations, and local communities must work together to implement policies that protect and restore wetland ecosystems.

Chapter 2: Wetlands and Climate Resilience

Wetlands play a crucial role in enhancing climate resilience by mitigating the impacts of extreme weather events, regulating water cycles, and sequestering carbon. As natural buffers, they reduce flood risks, protect coastlines from storm surges, and help maintain water availability during droughts. Their ability to absorb and store carbon also makes them valuable in climate change mitigation strategies.

Despite these benefits, wetlands are increasingly threatened by climate change, with rising temperatures, changing precipitation patterns, and sea-level rise altering their ecological balance. Protecting and restoring wetlands is essential for strengthening climate resilience at local and global scales.

This chapter explores the role of wetlands in climate regulation, highlighting their function as carbon sinks, flood mitigators, and drought buffers. It also examines the challenges posed by climate change and the importance of integrating wetland conservation into broader adaptation and mitigation strategies.

Carbon Sequestration

Wetlands play a crucial role in mitigating climate change by acting as significant carbon sinks. They store vast amounts of carbon in their soils and vegetation, reducing atmospheric carbon dioxide levels. Unlike many terrestrial ecosystems, wetlands can accumulate and retain carbon for thousands of years due to their unique hydrological and biological conditions. However, human activities and climate change threaten their ability to function as carbon reservoirs, leading to carbon release into the atmosphere. Understanding how wetlands sequester carbon and the factors that influence their carbon storage potential is essential for integrating wetland conservation into global climate strategies.

The Role of Wetlands in Carbon Sequestration

Wetlands capture carbon primarily through the accumulation of organic matter in waterlogged soils. Unlike upland environments, where decomposition is rapid, wetlands create anaerobic (low-oxygen) conditions that slow down microbial activity and organic matter breakdown. This allows plant material to accumulate as peat or organic-rich sediments, storing carbon over long periods.

- **Peatlands as Carbon Reservoirs**
  Peatlands, including bogs and fens, are the most effective wetland types for long-term carbon sequestration. These ecosystems store nearly one-third of the world's terrestrial carbon, despite covering only about 3% of the Earth's land surface. The slow decomposition rates in peatlands allow for continuous carbon accumulation, making them one of the most efficient carbon sinks on the planet. However, when peatlands are drained or degraded, they become major sources of carbon emissions, contributing to climate change.
- **Coastal Wetlands: Mangroves, Salt Marshes, and Seagrass Meadows**
  Coastal wetlands, including mangrove forests, salt marshes, and seagrass meadows, also serve as critical carbon sinks. Often referred to as "blue carbon" ecosystems, they store carbon both in their vegetation and in deep, water-saturated soils. Mangrove forests, in particular, can store up to four times more carbon per hectare than tropical rainforests. These ecosystems not only sequester carbon but also protect coastlines from erosion, storm surges, and sea-level rise.
- **Freshwater Wetlands and Floodplains**
  Inland freshwater wetlands, such as swamps and floodplains, also contribute to carbon storage by trapping organic material in sediments. While they may not store as much carbon per hectare as peatlands or mangroves, their widespread distribution makes them significant in regional and global carbon budgets. These wetlands help regulate greenhouse gas fluxes by balancing carbon sequestration and methane emissions.

Factors Affecting Carbon Sequestration in Wetlands

Several factors influence the carbon storage potential of wetlands, including hydrology, vegetation type, soil composition, and climate conditions.

- **Hydrology and Water Levels**
  Water saturation is essential for maintaining anaerobic conditions that slow organic matter decomposition. Changes in hydrology, such as drainage, water extraction, or prolonged droughts, can expose stored carbon to oxygen, accelerating decomposition and releasing carbon dioxide into the atmosphere. Maintaining stable water levels is crucial for preserving the carbon sequestration capacity of wetlands.
- **Vegetation Productivity and Decomposition Rates**
  The type and density of vegetation affect how much carbon a wetland can store. Highly productive wetlands with dense plant growth, such as mangrove forests and reed-dominated marshes, capture more carbon through photosynthesis. However, decomposition rates also play a role; wetlands with slow decomposition, such as peatlands, store carbon more effectively than wetlands where organic matter breaks down quickly.
- **Soil Composition and Peat Accumulation**
  Wetlands with organic-rich soils, particularly peatlands, are more effective at long-term carbon storage than those with mineral soils. Peat accumulation occurs when organic material is deposited faster than it decomposes, leading to thick layers of carbon-rich soil. Peatlands can store carbon for thousands of years, but disturbances such as drainage, agriculture, and peat extraction can release this stored carbon back into the atmosphere.
- **Climate Change and Greenhouse Gas Emissions**
  While wetlands are effective carbon sinks, they can also be sources of methane (CH_4), a potent greenhouse gas. The balance between carbon sequestration and methane emissions depends on factors such as water levels, temperature, and microbial activity. In general, peatlands and coastal wetlands

sequester more carbon than the amount of methane they release, making their conservation crucial for climate mitigation. However, rising temperatures and altered precipitation patterns due to climate change may shift this balance, increasing methane emissions and reducing carbon storage efficiency.

Threats to Wetland Carbon Sequestration

Human activities and environmental changes are reducing the ability of wetlands to function as carbon sinks. The primary threats include land-use changes, pollution, and climate-related disturbances.

- **Wetland Drainage and Land Conversion**
  The drainage of wetlands for agriculture, urban development, and infrastructure projects is one of the biggest threats to their carbon sequestration ability. When wetlands are drained, oxygen enters the soil, accelerating the decomposition of stored organic matter and releasing large amounts of carbon dioxide. Peatland drainage, in particular, is a major source of carbon emissions, with some drained peatlands emitting more carbon annually than entire countries.
- **Deforestation of Mangroves and Coastal Wetlands**
  The destruction of mangroves and other coastal wetlands for aquaculture, logging, and coastal development significantly reduces carbon storage capacity. Mangrove deforestation releases both aboveground biomass carbon and deep soil carbon, contributing to global greenhouse gas emissions. The loss of these ecosystems also reduces their ability to protect coastlines from climate impacts.
- **Pollution and Nutrient Overload**
  Excess nutrients from agricultural runoff and wastewater discharge can alter wetland ecosystems, leading to increased decomposition rates and greenhouse gas emissions. Eutrophication, caused by high nutrient levels, can shift wetland dynamics, reducing their ability to store carbon effectively. Contaminants such as heavy metals and

pesticides can also disrupt microbial communities that contribute to carbon sequestration.
- **Climate Change-Induced Impacts**
Rising temperatures, shifting rainfall patterns, and extreme weather events affect wetland carbon storage. Droughts can dry out wetlands, leading to oxidation of organic matter and carbon release. Increased frequency of storms and sea-level rise can erode coastal wetlands, reducing their ability to sequester carbon. Protecting wetlands from climate-driven disturbances is essential for maintaining their role as long-term carbon sinks.

The Importance of Protecting Wetlands for Climate Mitigation

Given their ability to capture and store carbon, wetlands are increasingly recognized as valuable assets in global climate mitigation efforts. Conservation, restoration, and sustainable management of wetlands can help maintain and enhance their carbon sequestration potential. International frameworks such as the Ramsar Convention, the UN Framework Convention on Climate Change (UNFCCC), and the Paris Agreement emphasize the importance of wetland protection in reducing greenhouse gas emissions.

- **Wetland Restoration for Carbon Sequestration**
Restoring degraded wetlands, such as rewetting drained peatlands and reforesting mangroves, can help recover lost carbon storage capacity. Rewetting peatlands prevents further carbon loss, while restoring mangroves enhances coastal resilience and carbon capture. Large-scale wetland restoration projects are being implemented globally to offset carbon emissions and improve climate resilience.
- **Integrating Wetlands into Climate Policies**
Many countries are incorporating wetland conservation into their climate action plans, recognizing their role in carbon sequestration. Wetland protection is being included in Nationally Determined Contributions (NDCs) under the Paris Agreement, and carbon markets are beginning to explore

mechanisms for financing wetland conservation through carbon credits.

Flood and Storm Protection

Wetlands play a crucial role in protecting communities and ecosystems from the impacts of floods and storms. Acting as natural buffers, they absorb and store excess water, slow runoff, and reduce the intensity of storm surges. As climate change increases the frequency and severity of extreme weather events, the importance of wetlands in flood and storm mitigation is becoming even more evident. However, wetland loss due to human activities threatens their ability to provide these essential services. Understanding how wetlands contribute to flood and storm protection is critical for integrating them into climate adaptation and disaster risk reduction strategies.

How Wetlands Reduce Flood Risks

Wetlands regulate water flow by acting as natural sponges, absorbing and storing excess rainfall and floodwaters. This process reduces the risk of downstream flooding by slowing water movement and gradually releasing it over time.

- **Water Absorption and Storage**
  Wetlands, particularly floodplain wetlands and marshes, can hold large volumes of water, reducing peak flood levels. During heavy rainfall or rapid snowmelt, they store excess water, preventing it from overwhelming rivers and urban drainage systems. This capacity is especially important in regions prone to seasonal flooding.
- **Flow Regulation and Delayed Runoff**
  By slowing down water flow, wetlands reduce the speed at which floodwaters reach populated areas. This prevents flash floods and erosion while allowing water to infiltrate into groundwater reserves. Floodplain wetlands along rivers help

moderate flood intensity by distributing water over a larger area, reducing damage to infrastructure and communities.

- **Sediment and Debris Capture**
Wetlands help trap sediments, nutrients, and debris carried by floodwaters, preventing them from accumulating in rivers and reservoirs. This improves water quality and reduces the risk of sediment buildup that can exacerbate future flooding. Vegetation in wetlands acts as a natural filter, stabilizing soil and reducing erosion.

Coastal Wetlands and Storm Surge Protection

Coastal wetlands, including mangroves, salt marshes, and seagrass meadows, serve as protective barriers against storm surges and extreme weather events. These wetlands dissipate wave energy, reducing the impact of hurricanes, typhoons, and cyclones on coastal communities.

- **Wave Energy Dissipation**
Mangrove forests and salt marshes absorb wave energy, decreasing wave height and strength before they reach shorelines. Studies have shown that mangroves can reduce wave heights by up to 66%, significantly lowering the risk of coastal flooding and erosion.
- **Storm Surge Reduction**
Coastal wetlands act as buffers, absorbing excess water from storm surges and reducing inland flooding. Salt marshes and mangroves slow the movement of water, allowing it to spread out and lose intensity before reaching developed areas. This function is critical in low-lying coastal regions vulnerable to rising sea levels and storm-driven flooding.
- **Erosion Control and Shoreline Stabilization**
Wetland vegetation stabilizes coastlines by binding soil and reducing the impact of wave action. Seagrass meadows and mangrove roots trap sediments, preventing shoreline retreat and land loss. This natural defense is particularly important for protecting infrastructure, homes, and agricultural land in coastal zones.

The Impact of Wetland Loss on Flood and Storm Resilience

Despite their effectiveness in flood and storm protection, wetlands are being lost at an alarming rate due to human activities. The degradation of wetlands reduces their capacity to absorb floodwaters and buffer storm impacts, increasing the vulnerability of communities to extreme weather events.

- **Urban Expansion and Drainage**
  Wetlands are often drained and converted for agriculture, urban development, and industrial use. This reduces their ability to store water and increases surface runoff, leading to higher flood risks. In many cities, natural wetlands have been replaced by impervious surfaces such as roads and buildings, which accelerate water flow and contribute to urban flooding.
- **Coastal Development and Mangrove Loss**
  Coastal wetlands are under threat from land reclamation, tourism development, and aquaculture expansion. The removal of mangroves and salt marshes weakens coastal defenses, making shorelines more vulnerable to storm surges and erosion. Without these natural barriers, storm impacts are more severe, leading to greater economic and human losses.
- **Climate Change and Rising Sea Levels**
  Climate change is altering wetland hydrology, affecting their ability to provide flood and storm protection. Rising sea levels can lead to saltwater intrusion in coastal wetlands, reducing their effectiveness as protective buffers. Increased frequency of extreme weather events, including hurricanes and heavy rainfall, puts additional stress on wetland ecosystems.

The Role of Wetlands in Climate Adaptation and Disaster Risk Reduction

Recognizing the importance of wetlands in flood and storm resilience, governments and organizations are integrating wetland conservation into climate adaptation and disaster risk management strategies. Protecting and restoring wetlands can reduce disaster risks and enhance resilience to climate-related hazards.

- **Wetland Restoration for Flood Management**
  Restoring degraded wetlands helps re-establish their water retention capacity, reducing flood risks in both urban and rural areas. Floodplain reconnection projects, where rivers are allowed to overflow into natural wetlands, provide a cost-effective alternative to engineered flood defenses. These projects have been successfully implemented in countries such as the Netherlands and the United States.
- **Integrating Wetlands into Coastal Protection Plans**
  Many coastal cities are incorporating wetland conservation into resilience planning. For example, mangrove reforestation projects are being promoted as natural solutions to storm protection in Southeast Asia and the Caribbean. These initiatives not only strengthen coastal defenses but also provide habitat for marine biodiversity and support local livelihoods.
- **Policy and Financial Support for Wetland Conservation**
  International agreements such as the Ramsar Convention and the Sendai Framework for Disaster Risk Reduction recognize wetlands as valuable assets for climate adaptation. Funding mechanisms, including payments for ecosystem services and carbon offset programs, are being developed to support wetland conservation and restoration efforts.

Challenges in Wetland Protection and Management

Despite growing awareness of the benefits of wetlands, challenges remain in their conservation and management. Competing land-use demands, weak enforcement of environmental regulations, and lack of public awareness hinder effective wetland protection.

- **Balancing Development and Conservation**
  In many regions, economic development pressures lead to wetland destruction. Striking a balance between

infrastructure expansion and wetland conservation requires integrated land-use planning and strong governance.
- **Legal and Institutional Barriers**
Many wetlands lack formal protection, making them vulnerable to degradation. Strengthening environmental policies and enforcing wetland protection laws are essential for long-term conservation.
- **Public Engagement and Awareness**
Increasing community participation in wetland conservation can enhance local resilience to flooding and storm risks. Educating stakeholders about the value of wetlands and promoting sustainable land management practices are key steps in preserving these ecosystems.

Drought Resilience

Wetlands play a critical role in maintaining water availability and mitigating the impacts of droughts. By acting as natural reservoirs, they store and release water gradually, helping to sustain river flows, groundwater levels, and ecosystems during dry periods. Wetlands also support biodiversity and provide essential ecosystem services that contribute to water security for both natural and human systems. However, climate change, land-use changes, and increasing water demand are threatening their ability to function as drought buffers. Protecting and restoring wetlands is essential for enhancing drought resilience and ensuring long-term water sustainability.

How Wetlands Contribute to Drought Resilience

Wetlands enhance drought resilience by regulating hydrological cycles, recharging groundwater, and maintaining ecological stability in arid and semi-arid regions. Their ability to store water during wet periods and release it slowly over time makes them essential for sustaining freshwater availability.

Water Storage and Slow Release

Wetlands act as natural water storage systems, accumulating water from rainfall, surface runoff, and groundwater inflows. During periods of drought, they gradually release stored water, helping to maintain streamflow and prevent water shortages.

- **Seasonal Water Regulation**
  Many wetlands experience seasonal fluctuations in water levels, allowing them to store water during wet seasons and release it when conditions become dry. This function is particularly important in river basins and floodplain wetlands, where seasonal droughts are common.
- **Reducing Evaporation Losses**
  Compared to artificial reservoirs, wetlands lose less water through evaporation due to the presence of dense vegetation, which provides shade and reduces water loss. This makes them more efficient at sustaining water supplies during prolonged dry spells.

Groundwater Recharge and Retention

Wetlands play a key role in replenishing groundwater reserves by allowing water to percolate into underground aquifers. This helps maintain water availability for agriculture, drinking water supplies, and ecosystems, particularly in regions that rely on groundwater during droughts.

- **Enhancing Groundwater Recharge**
  Wetlands, especially floodplains and marshes, facilitate the slow infiltration of water into the ground, helping to recharge depleted aquifers. This process ensures a steady supply of water even when surface sources dry up.
- **Preventing Groundwater Depletion**
  In areas where wetlands have been drained or degraded, groundwater depletion accelerates, increasing the risk of water shortages during droughts. Protecting wetlands helps stabilize groundwater levels and reduce the need for excessive extraction.

Biodiversity and Ecosystem Stability During Droughts

Wetlands provide habitat for a wide range of species that depend on stable water conditions, making them crucial for maintaining biodiversity during drought periods. Many aquatic and semi-aquatic species rely on wetlands as refuges when other water sources dry up.

- **Supporting Migratory and Resident Species**
  Wetlands serve as critical habitats for migratory birds, amphibians, and fish that require stable water sources during dry seasons. They provide breeding, feeding, and sheltering grounds for wildlife, ensuring the survival of species adapted to fluctuating water conditions.
- **Maintaining Ecosystem Services**
  Even during droughts, wetlands continue to support essential ecological processes, such as nutrient cycling, water filtration, and soil stabilization. This resilience helps prevent ecosystem collapse and ensures that wetlands can recover quickly once water levels are restored.

Threats to Wetland Resilience During Droughts

Despite their ability to buffer against droughts, wetlands are increasingly under threat from human activities and climate change. These threats reduce their capacity to store water, recharge groundwater, and sustain biodiversity.

Wetland Drainage and Land Conversion

One of the biggest threats to wetland resilience is land conversion for agriculture, urban expansion, and infrastructure development. Draining wetlands for farming or construction disrupts their natural water storage function, reducing their ability to mitigate drought impacts.

- **Agricultural Expansion**
  Many wetlands have been drained to create farmland, resulting in significant water loss and reduced drought resilience. Without wetlands to store and release water, agricultural areas become more vulnerable to dry conditions.
- **Urbanization and Infrastructure Development**
  Urban expansion replaces wetlands with impermeable surfaces, such as roads and buildings, which prevent water infiltration and increase surface runoff. This leads to reduced groundwater recharge and heightened vulnerability to droughts.

Overextraction of Water Resources

Excessive water withdrawal for irrigation, industry, and domestic use puts additional stress on wetlands, reducing their capacity to store and release water during droughts.

- **Groundwater Overuse**
  In many regions, wetlands rely on groundwater inflows to maintain water levels. Overpumping of groundwater for agriculture and urban supply lowers water tables, causing wetlands to dry out and lose their drought-buffering function.
- **Surface Water Diversions**
  Dams, irrigation channels, and river modifications disrupt natural water flows, depriving wetlands of their primary water sources. This leads to habitat degradation and reduced water availability during drought periods.

Climate Change and Altered Hydrological Patterns

Climate change is intensifying drought conditions in many parts of the world, altering precipitation patterns and increasing temperatures. These changes affect wetland hydrology, making it harder for them to function as effective drought buffers.

- **Reduced Rainfall and Longer Drought Periods**
  Many wetlands depend on seasonal rainfall to maintain water levels. Declining precipitation and prolonged dry spells reduce their ability to store and release water, leading to increased drying and ecosystem stress.
- **Increased Evapotranspiration**
  Rising temperatures accelerate water loss from wetlands through evapotranspiration, further reducing their ability to sustain water supplies during droughts. This effect is particularly pronounced in shallow wetlands and those in arid regions.

Strategies for Enhancing Wetland Drought Resilience

Protecting and restoring wetlands is essential for maintaining their role in mitigating droughts. Sustainable management practices can enhance their resilience and ensure long-term water security.

Wetland Restoration and Conservation

Restoring degraded wetlands helps re-establish their natural hydrological functions, improving their ability to store water and support biodiversity during droughts.

- **Rewetting Drained Wetlands**
  Reintroducing water to previously drained wetlands helps restore their water retention capacity and enhances groundwater recharge. This process is being implemented in several wetland conservation projects worldwide.
- **Protecting Existing Wetlands**
  Strengthening legal protections and enforcing conservation policies prevent further wetland degradation, ensuring they continue to function as drought buffers.

Sustainable Water Management

Integrating wetlands into broader water management strategies can enhance regional drought resilience and support long-term water sustainability.

- **Maintaining Natural Water Flows**
  Ensuring that wetlands receive adequate water through regulated flow management and water allocation policies can help sustain their ecological functions during droughts.
- **Reducing Overextraction**
  Implementing sustainable groundwater and surface water use policies prevents excessive depletion of wetland water sources, maintaining their resilience in dry periods.

Community Involvement and Policy Integration

Engaging local communities and policymakers in wetland conservation efforts strengthens long-term resilience and promotes sustainable land and water use.

- **Promoting Wetland-Based Agriculture**
  Encouraging farming practices that maintain wetland hydrology, such as agroforestry and controlled grazing, helps balance economic development with conservation.
- **Integrating Wetlands into Drought Adaptation Plans**
  Governments and environmental organizations are increasingly recognizing wetlands as key assets in climate adaptation strategies. Including wetland conservation in national drought policies can enhance resilience and support water security.

Chapter 3: Wetlands and Biodiversity Conservation

Wetlands are among the most biologically diverse ecosystems on Earth, providing habitat for a vast array of plant, animal, and microbial species. They support both aquatic and terrestrial life, serving as breeding, feeding, and migration stopover points for numerous species, including fish, amphibians, birds, and mammals. Their rich biodiversity contributes to ecosystem stability, nutrient cycling, and food security.

However, wetland biodiversity is increasingly under threat from habitat loss, pollution, invasive species, and climate change. Many wetland-dependent species are facing population declines or extinction due to human activities that disrupt their delicate ecological balance.

This chapter explores the critical role of wetlands in supporting biodiversity, highlighting the complex interdependencies between species and their habitats. It examines the threats to wetland biodiversity and discusses conservation strategies to protect and restore these vital ecosystems. Understanding the importance of wetlands for biodiversity is essential for developing effective conservation policies and ensuring the long-term survival of wetland-dependent species.

Biodiversity Hotspots

Wetlands are among the most biologically diverse ecosystems on Earth, supporting an incredible range of species across different habitats. From coastal mangroves to inland freshwater marshes, wetlands provide essential breeding, feeding, and sheltering grounds for countless organisms. Many wetlands are considered biodiversity hotspots due to their high species richness and ecological importance. These areas are crucial for global biodiversity conservation, yet they are increasingly threatened by habitat destruction, pollution, and climate change.

Why Wetlands Are Biodiversity Hotspots

Wetlands support high biodiversity due to their unique ecological characteristics. Their dynamic nature, fluctuating water levels, and rich nutrient cycles create a variety of microhabitats that sustain diverse plant and animal life. Key factors contributing to their biodiversity include:

- **Varied Habitat Types**: Wetlands encompass multiple environments, including open water, mudflats, and vegetated areas, providing diverse ecological niches.
- **High Primary Productivity**: Wetlands are highly productive ecosystems, supporting complex food webs that sustain a wide range of species.
- **Connectivity Between Ecosystems**: Many wetlands act as transition zones between terrestrial and aquatic environments, allowing species movement and genetic exchange.
- **Critical Role in Life Cycles**: Wetlands serve as breeding grounds for amphibians, fish, and birds, making them essential for species survival.

Major Wetland Biodiversity Hotspots

Several wetland regions around the world are recognized for their exceptional biodiversity. These areas host numerous endemic and migratory species, playing a crucial role in global biodiversity conservation.

The Amazon Wetlands (South America)

The Amazon Basin contains extensive wetlands, including floodplain forests and oxbow lakes, which are among the most biodiverse in the world. These wetlands support an extraordinary range of species, including:

- **Mammals**: Jaguars, giant otters, and manatees rely on the Amazon's wetland ecosystems for survival.
- **Birds**: The region is home to thousands of bird species, including macaws and herons.
- **Fish**: Wetland waters host over 3,000 species of fish, including the famous arapaima and piranha.

However, deforestation, hydroelectric dams, and agricultural expansion threaten the integrity of these wetlands, impacting both biodiversity and local communities.

The Okavango Delta (Africa)

Located in Botswana, the Okavango Delta is a vast inland wetland that supports a remarkable diversity of species. Seasonal flooding transforms this arid region into a lush ecosystem, attracting wildlife from across Africa. Key species include:

- **Large Mammals**: Elephants, lions, leopards, and hippos depend on the delta's water sources.
- **Birdlife**: Over 400 bird species, including African fish eagles and kingfishers, thrive in the wetland's diverse habitats.
- **Fish and Amphibians**: The delta's seasonal floods sustain fish populations that provide food for birds, reptiles, and local communities.

Despite its designation as a UNESCO World Heritage Site, the Okavango Delta faces threats from climate change, water extraction, and infrastructure development.

The Sundarbans (Asia)

The Sundarbans, the world's largest mangrove forest, spans Bangladesh and India, providing a crucial habitat for many endangered species. This wetland ecosystem is known for:

- **The Bengal Tiger**: The Sundarbans are home to one of the last remaining populations of the endangered Bengal tiger.
- **Diverse Aquatic Life**: Dolphins, saltwater crocodiles, and mudskippers thrive in the delta's brackish waters.
- **Rich Birdlife**: Migratory shorebirds and wading birds rely on the region for feeding and breeding.

The Sundarbans are threatened by rising sea levels, deforestation, and pollution, posing risks to both wildlife and the millions of people who depend on its resources.

The Everglades (North America)

The Florida Everglades is a unique subtropical wetland known for its slow-moving waters and diverse habitats. It supports:

- **Endangered Species**: The American alligator, Florida panther, and manatee all depend on the Everglades.
- **Wading Birds**: The wetland is an important nesting site for herons, egrets, and ibises.
- **Freshwater Fish and Amphibians**: The Everglades support a variety of aquatic species that form the base of the wetland's food web.

Urban expansion, water diversion, and invasive species threaten this ecosystem, making conservation efforts crucial for maintaining its biodiversity.

The Pantanal (South America)

The Pantanal, spanning Brazil, Bolivia, and Paraguay, is the world's largest tropical wetland, covering nearly 200,000 square kilometers. It is home to:

- **The World's Largest Jaguar Population**: This wetland is a stronghold for jaguars, offering abundant prey and protected habitat.
- **Giant Otters and Tapirs**: These iconic species rely on the Pantanal's waterways.
- **Vast Bird Populations**: The wetland attracts thousands of migratory birds, including the Jabiru stork.

Despite its ecological richness, the Pantanal faces challenges from deforestation, wildfires, and agricultural expansion.

Threats to Wetland Biodiversity Hotspots

While wetlands provide essential habitat for countless species, they are also among the most threatened ecosystems globally. Major threats to wetland biodiversity hotspots include:

- **Habitat Destruction**: Urbanization, deforestation, and land reclamation are leading to significant wetland loss.
- **Pollution**: Agricultural runoff, industrial waste, and plastic pollution degrade water quality and harm aquatic life.
- **Climate Change**: Rising temperatures, altered rainfall patterns, and sea-level rise threaten wetland ecosystems.
- **Overexploitation**: Overfishing, illegal wildlife trade, and excessive resource extraction put pressure on wetland species.

The Importance of Conserving Wetland Biodiversity Hotspots

Protecting wetlands is essential for maintaining global biodiversity and ecological balance. Conservation efforts focus on:

- **Establishing Protected Areas**: Designating wetlands as national parks, Ramsar sites, and UNESCO World Heritage Sites helps safeguard their biodiversity.
- **Sustainable Management**: Integrating wetlands into climate adaptation and land-use planning ensures long-term conservation.
- **Community Involvement**: Engaging local communities in wetland conservation promotes sustainable resource use and enhances biodiversity protection.

Species Interdependence

Wetlands are complex ecosystems where species interact in intricate and interdependent ways. These interconnections sustain biodiversity, regulate ecological processes, and maintain wetland functions such as water purification, flood control, and carbon sequestration. Every organism in a wetland ecosystem, from microorganisms to apex predators, plays a role in sustaining the system's balance. Disruptions to these relationships, whether from habitat loss, pollution, or climate change, can lead to cascading effects that alter the entire ecosystem. Understanding species interdependence in wetlands is essential for effective conservation and ecosystem management.

Trophic Relationships and Food Webs

Wetland ecosystems support highly dynamic food webs, with species playing roles as primary producers, herbivores, predators, and decomposers. These trophic interactions regulate population dynamics and ensure nutrient cycling within the ecosystem.

Primary Producers: The Foundation of Wetland Life

Plants, algae, and phytoplankton serve as primary producers, converting sunlight into energy through photosynthesis. These organisms form the base of the food web, supporting a wide range of herbivores and omnivores.

- Emergent vegetation, such as reeds, cattails, and sedges, provides food and shelter for insects, birds, and amphibians.
- Submerged plants like eelgrass and pondweed oxygenate the water and support aquatic herbivores.
- Algae and phytoplankton sustain invertebrates and small fish, which in turn feed larger organisms.

Without primary producers, wetland ecosystems would collapse, as they supply the essential energy needed for all higher trophic levels.

Herbivores and Omnivores: Key Consumers

Herbivores in wetlands play a critical role in maintaining plant diversity and transferring energy up the food chain. They consume plant material, control vegetation growth, and serve as prey for predators.

- Insects, such as dragonfly larvae and water beetles, feed on aquatic plants and algae while serving as an important food source for fish and birds.
- Fish species, including carp and tilapia, graze on submerged vegetation and contribute to nutrient cycling.
- Amphibians, such as frogs and salamanders, consume aquatic plants and invertebrates while acting as both predators and prey.
- Waterfowl, like ducks and geese, rely on wetland plants for food while dispersing seeds through their movements.

Omnivores, which consume both plant and animal matter, help stabilize the ecosystem by adapting to seasonal food availability. Raccoons, turtles, and certain bird species shift their diets based on resource abundance, ensuring continuity in nutrient flow.

Predators: Regulating Population Balance

Predators control herbivore populations, preventing overgrazing and maintaining ecosystem stability. Wetlands support a diverse range of carnivores, including fish, reptiles, birds, and mammals.

- Fish predators, such as bass and pike, regulate populations of smaller fish and invertebrates, maintaining a balanced aquatic ecosystem.
- Birds of prey, including herons, ospreys, and eagles, hunt fish, amphibians, and small mammals, preventing overpopulation.
- Reptilian predators, such as alligators and crocodiles, help control wetland food chains by feeding on fish, birds, and mammals.
- Mammalian predators, like otters and foxes, manage populations of fish, rodents, and birds, contributing to biodiversity balance.

Without predator species, herbivore populations could grow unchecked, leading to vegetation loss and habitat degradation.

Mutualism and Symbiotic Relationships

Wetland species often form mutually beneficial relationships, enhancing survival and ecological resilience. These interactions help sustain biodiversity, ensure pollination, and facilitate nutrient exchange.

Pollination and Seed Dispersal

Many wetland plants rely on insects and birds for pollination, ensuring reproductive success and genetic diversity.

- Bees, butterflies, and beetles pollinate flowering wetland plants, aiding in seed production.

- Hummingbirds and nectar-feeding bats play a crucial role in tropical wetland ecosystems by pollinating specialized plant species.
- Waterfowl and mammals disperse seeds by consuming fruits and carrying them to new locations, promoting plant colonization.

This interdependence ensures wetland vegetation remains diverse and capable of adapting to environmental changes.

Microbial and Plant Symbiosis

Wetlands host diverse microbial communities that support plant health and nutrient cycling.

- Nitrogen-fixing bacteria, such as those in the roots of wetland plants like alders and legumes, enhance soil fertility by converting atmospheric nitrogen into a usable form.
- Mycorrhizal fungi form partnerships with plant roots, improving water and nutrient absorption while receiving sugars in return.
- Decomposers, including fungi and bacteria, break down organic material, recycling nutrients essential for plant growth.

These interactions enhance wetland productivity and resilience, ensuring sustained ecosystem functions.

Keystone Species and Ecosystem Engineers

Some wetland species have a disproportionate impact on their environment, shaping habitat structure and influencing the survival of other species.

Beavers: Wetland Architects

Beavers are well-known ecosystem engineers, altering landscapes by building dams and creating new wetland habitats. Their activities:

- Slow water flow, reducing erosion and increasing groundwater recharge.
- Create ponds that provide habitat for fish, amphibians, and waterfowl.
- Increase plant diversity by allowing new wetland vegetation to establish.

By modifying wetland hydrology, beavers enhance ecosystem complexity and biodiversity.

Mangroves: Coastal Protectors

Mangrove trees stabilize shorelines, reduce storm surges, and provide essential habitat for marine life. Their complex root systems:

- Trap sediments, improving water quality.
- Offer shelter for juvenile fish and crustaceans.
- Support diverse bird and mammal populations.

Without mangroves, many coastal wetlands would face increased erosion and biodiversity loss.

Alligators and Crocodiles: Apex Predators

As top predators, alligators and crocodiles regulate wetland food chains. They:

- Control populations of fish, turtles, and mammals.
- Create "gator holes" during dry periods, providing refuge for aquatic species.
- Maintain balance among prey populations, preventing overgrazing and habitat destruction.

Their presence ensures wetland ecosystems remain healthy and diverse.

Threats to Species Interdependence in Wetlands

Human activities are disrupting the delicate balance of species interactions in wetlands, leading to biodiversity loss and ecosystem instability. Major threats include:

- **Habitat destruction**: Wetland drainage, land conversion, and infrastructure development reduce available habitat for wetland-dependent species.
- **Pollution**: Agricultural runoff, industrial waste, and chemical contaminants harm wetland species and disrupt food webs.
- **Climate change**: Rising temperatures, altered precipitation patterns, and sea-level rise threaten wetland ecosystems and species survival.
- **Invasive species**: Non-native plants and animals outcompete native species, disrupting ecological relationships and altering wetland dynamics.

Threats to Biodiversity

Wetlands are among the most biologically diverse ecosystems on Earth, providing habitat for countless species of plants, animals, and microorganisms. However, they are also among the most threatened ecosystems, with biodiversity declining at an alarming rate due to human activities and environmental changes. The degradation and loss of wetlands disrupt ecological balance, reduce species populations, and weaken the ability of these ecosystems to provide essential services such as water purification, carbon storage, and flood protection. Understanding the key threats to wetland biodiversity is crucial for developing effective conservation strategies and ensuring the long-term sustainability of these vital ecosystems.

Habitat Destruction and Wetland Loss

The most significant threat to wetland biodiversity is habitat destruction, often caused by land conversion, drainage, and infrastructure development. Wetlands are frequently considered "wastelands" and are drained or filled to make way for agriculture, urban expansion, and industrial projects.

- **Agricultural Expansion**
  Agriculture is one of the primary drivers of wetland loss. Wetlands are often drained to create farmland, as their nutrient-rich soils support high crop yields. However, this destroys natural habitats, leading to declines in wetland-dependent species. The use of fertilizers and pesticides in agricultural areas also pollutes wetland water sources, further degrading biodiversity.
- **Urbanization and Infrastructure Development**
  Expanding cities, roads, and industrial areas encroach on wetland habitats. The construction of dams, levees, and drainage systems alters water flow patterns, disrupting the natural hydrological balance of wetlands. As urban areas grow, wetlands are increasingly being converted into residential and commercial developments, leading to habitat fragmentation and species displacement.
- **Aquaculture and Coastal Development**
  In coastal regions, wetlands such as mangroves and salt marshes are cleared for shrimp farms, fishponds, and tourism infrastructure. This results in the loss of breeding and feeding grounds for fish, shellfish, and migratory birds. Coastal wetland destruction also weakens natural defenses against storm surges and sea-level rise.

Pollution and Contamination

Pollution is a major threat to wetland biodiversity, affecting water quality and harming aquatic life. Pollutants enter wetlands through agricultural runoff, industrial discharge, and urban waste, leading to toxic conditions that disrupt ecological processes.

- **Agricultural Runoff and Eutrophication**
  Excess fertilizers and pesticides from agricultural fields wash into wetlands, introducing high levels of nitrogen and phosphorus. This leads to eutrophication, where excessive nutrient levels cause algal blooms. Algal blooms deplete oxygen levels in the water, creating "dead zones" where fish and other aquatic organisms cannot survive.
- **Industrial and Chemical Pollution**
  Factories, mining operations, and oil refineries release heavy metals, toxins, and hazardous chemicals into wetland ecosystems. These pollutants accumulate in the food chain, affecting fish, birds, and mammals. Mercury, lead, and cadmium are particularly harmful, causing developmental and reproductive issues in wetland species.
- **Plastic and Microplastic Contamination**
  Plastics are a growing problem in wetland environments, particularly in coastal and freshwater ecosystems. Discarded plastic waste entangles wildlife, while microplastics—tiny plastic particles—are ingested by fish, birds, and invertebrates, causing internal damage and bioaccumulation of toxins in the food web.

Climate Change and Extreme Weather Events

Climate change is altering wetland ecosystems by affecting temperature, precipitation, and sea levels. These changes impact species distributions, disrupt breeding cycles, and increase the frequency of extreme weather events.

- **Rising Temperatures and Altered Hydrological Cycles**
  Increasing global temperatures affect wetland hydrology by altering rainfall patterns and evaporation rates. Many wetlands rely on seasonal water flows, but changes in precipitation reduce water availability, leading to prolonged droughts in some regions and excessive flooding in others. These changes disrupt breeding and migration patterns for wetland-dependent species.

- **Sea-Level Rise and Coastal Wetland Loss**
  Rising sea levels threaten coastal wetlands, particularly mangroves and salt marshes. As seawater encroaches, it alters salinity levels and submerges wetland habitats, forcing species to migrate or adapt. In many cases, wetlands are unable to migrate inland due to human infrastructure, leading to permanent habitat loss.
- **Extreme Weather Events**
  More frequent and intense hurricanes, cyclones, and heavy storms damage wetland ecosystems by eroding shorelines, uprooting vegetation, and destroying breeding grounds for wildlife. Mangroves and marshes, which act as natural buffers against storms, are particularly vulnerable when their root systems are weakened by environmental stressors.

Invasive Species and Habitat Degradation

Invasive species pose a serious threat to wetland biodiversity by outcompeting native species, altering food webs, and disrupting ecological balance. Non-native plants, animals, and pathogens can rapidly spread in wetland environments, often thriving in disturbed habitats.

- **Invasive Plants**
  Non-native plant species, such as common reed (Phragmites australis), water hyacinth (Eichhornia crassipes), and purple loosestrife (Lythrum salicaria), aggressively spread in wetlands, crowding out native vegetation and altering habitat structure. These plants can reduce food and shelter availability for wetland wildlife, leading to population declines.
- **Invasive Animals**
  Introduced fish, amphibians, and mammals can disrupt native wetland ecosystems. For example, invasive fish species such as common carp disturb sediments and reduce water quality, making it difficult for native aquatic plants and invertebrates to survive. Non-native predators, such as cane toads and feral

cats, prey on native wetland species, leading to biodiversity loss.

- **Disease and Pathogens**
  Invasive pathogens, including fungi, bacteria, and viruses, can spread rapidly in wetland ecosystems, affecting amphibians, fish, and bird populations. The chytrid fungus (Batrachochytrium dendrobatidis) has decimated amphibian populations worldwide, while avian botulism outbreaks have killed thousands of waterfowl in wetlands.

Overexploitation of Wetland Resources

Human activities, including overfishing, logging, and unsustainable harvesting, place significant pressure on wetland biodiversity. Many wetland-dependent species are declining due to excessive resource extraction.

- **Overfishing and Habitat Degradation**
  Wetlands support vital fisheries, but overfishing reduces fish populations, disrupts food webs, and impacts local communities that depend on fishing for livelihoods. The destruction of spawning and nursery habitats further exacerbates declines in fish stocks.
- **Deforestation and Peat Extraction**
  Mangrove forests and peatlands are frequently harvested for timber, fuel, and commercial products. Peat extraction for agriculture and horticulture releases large amounts of carbon dioxide into the atmosphere while degrading wetland habitats for species that rely on them.
- **Illegal Wildlife Trade**
  Many wetland species, including turtles, birds, and amphibians, are captured and sold in illegal wildlife markets. The removal of these species from their natural habitats contributes to population declines and disrupts ecological interactions.

Chapter 4: Wetlands in Sustainable Water Management

Wetlands play a vital role in maintaining water quality, regulating water flows, and supporting freshwater availability. As natural water filters, they remove pollutants, trap sediments, and regulate nutrient levels, improving overall water quality. Wetlands also contribute to groundwater recharge, helping to sustain water supplies for agriculture, industry, and human consumption. Additionally, they act as buffers against extreme hydrological events, reducing the impacts of floods and droughts.

However, increasing water demand, pollution, and wetland degradation threaten their ability to provide these essential services. Unsustainable land-use practices, excessive water extraction, and climate change are putting pressure on wetland ecosystems worldwide.

This chapter explores the role of wetlands in sustainable water management, highlighting their function in water filtration, groundwater recharge, and balancing competing water demands. It also examines the challenges wetlands face and the need for integrated policies to ensure their conservation while supporting water security for future generations.

Water Filtration and Quality Improvement

Wetlands play a critical role in maintaining water quality by acting as natural filtration systems. They remove pollutants, trap sediments, and regulate nutrient levels, improving the overall health of aquatic ecosystems. Wetland plants, soils, and microbial communities work together to break down and absorb contaminants, making wetlands essential for sustaining clean water supplies for human use, agriculture, and biodiversity. However, increasing pollution, habitat destruction, and climate change threaten their ability to provide these services. Protecting and restoring wetlands is vital for ensuring long-term water quality and ecosystem health.

How Wetlands Improve Water Quality

Wetlands filter water through a combination of biological, chemical, and physical processes. These natural filtration mechanisms help remove harmful substances, including excess nutrients, heavy metals, and pathogens.

Sediment Trapping and Reduction of Turbidity

Wetlands slow down the flow of water, allowing sediments and suspended particles to settle. This reduces turbidity and prevents excessive sediment buildup in rivers, lakes, and reservoirs.

- **Vegetation as a Natural Barrier**
  Wetland plants, such as reeds and cattails, slow water movement and trap sediments, preventing soil erosion and reducing downstream sediment loads.
- **Retention of Particulate Matter**
  Fine sediments suspended in water carry pollutants such as heavy metals and organic matter. By trapping these particles, wetlands prevent contaminants from reaching larger water bodies.

Nutrient Absorption and Pollution Control

Wetlands regulate nutrient levels by absorbing excess nitrogen and phosphorus from agricultural runoff, wastewater, and stormwater. These nutrients, if unchecked, can lead to eutrophication—an overgrowth of algae that depletes oxygen and harms aquatic life.

- **Nitrogen Removal**
  Wetlands remove nitrogen through microbial processes such

as denitrification, where bacteria convert nitrates into nitrogen gas, releasing it harmlessly into the atmosphere.

- **Phosphorus Retention**
Wetland soils and vegetation absorb phosphorus, preventing it from accumulating in waterways and causing algal blooms. Over time, phosphorus is stored in sediments or taken up by plants.

Breakdown of Organic Pollutants and Pathogen Removal

Wetlands degrade and neutralize organic pollutants from wastewater and industrial discharge. Microbial communities break down harmful substances, improving water quality.

- **Microbial Decomposition**
Bacteria and fungi in wetland soils break down organic compounds, including pesticides, herbicides, and hydrocarbons from industrial pollution. This reduces the toxicity of contaminated water.
- **Pathogen Reduction**
Wetlands remove harmful bacteria, viruses, and parasites by exposing them to natural disinfection processes such as ultraviolet (UV) light, sedimentation, and microbial consumption. This helps improve public health by reducing waterborne diseases.

Types of Wetlands Used for Water Filtration

Different wetland types provide varying levels of water filtration, depending on their vegetation, hydrology, and soil composition.

Natural Wetlands

- **Marshes**
These wetlands, dominated by herbaceous plants, effectively remove sediments and nutrients from surface water, making

them valuable for improving agricultural runoff and urban stormwater quality.

- **Swamps**
  Forested wetlands provide filtration benefits by trapping pollutants in soil and slowing water flow, allowing for greater microbial decomposition.
- **Peatlands**
  Peat bogs and fens absorb and retain heavy metals and organic pollutants, acting as long-term carbon and contaminant storage systems.

Constructed Wetlands

- **Surface Flow Wetlands**
  These artificial wetlands mimic natural processes by using open-water areas with emergent vegetation to filter contaminants from wastewater and stormwater.
- **Subsurface Flow Wetlands**
  Water flows through gravel or soil substrates planted with wetland vegetation, allowing microbes to break down pollutants underground. These systems are often used for treating municipal and industrial wastewater.

Challenges to Wetland Water Filtration Capacity

Despite their effectiveness, wetlands face increasing threats that reduce their ability to filter pollutants and maintain water quality.

Pollution Overload

Wetlands can become overwhelmed when exposed to excessive pollution levels. High concentrations of nutrients, chemicals, and sediments can reduce their filtration efficiency and lead to ecosystem degradation.

- **Agricultural Runoff**
  Excessive fertilizers and pesticides entering wetlands can

exceed their natural capacity to absorb nutrients, leading to harmful algal blooms and biodiversity loss.

- **Industrial and Urban Wastewater**
  Untreated or poorly managed wastewater introduces heavy metals, pharmaceuticals, and microplastics, which wetlands struggle to break down fully.

Wetland Loss and Degradation

The destruction of wetlands for agriculture, urban development, and infrastructure projects reduces their ability to improve water quality. Wetland loss leads to:

- **Increased Pollution in Waterways**
  Without wetlands to filter contaminants, pollutants enter rivers and lakes, affecting drinking water supplies and aquatic ecosystems.
- **Declining Groundwater Quality**
  Wetlands contribute to groundwater recharge and filtration. Their loss reduces natural water purification, leading to contamination of underground water reserves.

Climate Change and Altered Hydrological Cycles

Climate change is affecting wetland water filtration functions by altering rainfall patterns, increasing droughts, and intensifying extreme weather events.

- **Drought and Reduced Water Levels**
  Lower water levels reduce wetland filtration capacity, leading to increased pollutant concentrations.
- **Flooding and Sediment Overload**
  Intense storms can wash excessive sediments and pollutants into wetlands, overwhelming their ability to filter water effectively.

Strategies for Protecting and Enhancing Wetland Filtration Functions

To ensure wetlands continue to improve water quality, conservation and restoration efforts must focus on maintaining their ecological integrity and hydrological balance.

Wetland Conservation and Restoration

- **Protecting Existing Wetlands**
  Establishing protected areas and enforcing environmental regulations prevent wetland loss and degradation.
- **Restoring Degraded Wetlands**
  Rewetting drained wetlands, removing invasive species, and reintroducing native vegetation can enhance filtration capacity and ecosystem resilience.

Sustainable Land and Water Management

- **Reducing Agricultural Runoff**
  Implementing buffer zones, cover crops, and precision fertilizer application reduces nutrient pollution entering wetlands.
- **Improving Wastewater Treatment**
  Expanding the use of constructed wetlands for treating municipal and industrial wastewater provides a cost-effective, natural filtration solution.

Climate Adaptation Strategies

- **Enhancing Wetland Resilience**
  Managing water flows, restoring natural hydrology, and increasing connectivity between wetlands and river systems help wetlands withstand climate-related challenges.
- **Incorporating Wetlands into Water Security Planning**
  Governments and communities can integrate wetland

conservation into broader water management policies to
ensure long-term water quality benefits.

Groundwater Recharge

Wetlands play a vital role in replenishing groundwater supplies by
allowing water to slowly infiltrate the soil and reach underground
aquifers. This process, known as groundwater recharge, is essential
for maintaining freshwater availability, supporting ecosystems, and
ensuring sustainable water use for agriculture, industry, and human
consumption. Wetlands act as natural sponges, capturing rainfall,
surface runoff, and floodwaters before gradually releasing the water
into the ground. However, increasing land-use changes, pollution,
and climate variability threaten their ability to provide this essential
service. Protecting and restoring wetlands is critical for maintaining
groundwater recharge and ensuring long-term water security.

How Wetlands Facilitate Groundwater Recharge

Wetlands contribute to groundwater recharge through a combination
of hydrological and ecological processes that allow water to move
from the surface into underground storage systems.

Water Infiltration and Percolation

Wetlands promote the slow infiltration of water into the soil,
reducing surface runoff and allowing moisture to penetrate deep into
the ground. This process is influenced by several factors:

- **Soil Composition**
  Wetlands with porous soils, such as sandy or loamy soils,
  facilitate faster infiltration and higher groundwater recharge
  rates. In contrast, clay-rich soils absorb water more slowly,
  leading to surface retention before gradual percolation.
- **Vegetation and Root Systems**
  Wetland plants enhance groundwater recharge by stabilizing
  soil and promoting water movement through root channels.

The deep roots of certain species, such as reeds and sedges, help water penetrate deeper layers of the soil, improving infiltration efficiency.

- **Water Table Connectivity**
Some wetlands are directly connected to underlying aquifers, allowing water to seep through soil layers and recharge groundwater supplies. Floodplain wetlands and riparian zones near rivers play a key role in maintaining groundwater levels in these interconnected systems.

Regulation of Groundwater Flow

Wetlands not only recharge groundwater but also help regulate the flow of water within aquifers. By slowly releasing stored water, they prevent excessive depletion and maintain stable water levels during dry periods.

- **Seasonal Recharge Cycles**
Many wetlands recharge groundwater at different rates throughout the year, depending on seasonal rainfall and hydrological conditions. During wet seasons, wetlands absorb and store excess water, which is gradually released into underground aquifers when conditions become dry.
- **Buffering Against Groundwater Depletion**
In regions where groundwater is heavily extracted for irrigation, drinking water, and industrial use, wetlands act as a buffer by replenishing aquifers and reducing the risk of water shortages. This function is particularly important in arid and semi-arid landscapes where groundwater is the primary water source.

Types of Wetlands That Contribute to Groundwater Recharge

Different types of wetlands vary in their ability to recharge groundwater, depending on their hydrology, location, and soil structure.

Floodplain Wetlands

Floodplains adjacent to rivers and streams play a significant role in groundwater recharge by storing excess water during floods and gradually releasing it into surrounding soils. These wetlands:

- Absorb large volumes of water during heavy rainfall and river overflows.
- Slowly filter water into underground aquifers, maintaining base flows in rivers.
- Support biodiversity by sustaining water availability during dry periods.

Marshes and Swamps

Inland wetlands, including freshwater marshes and swamps, contribute to groundwater recharge by allowing standing water to infiltrate the soil. These wetlands:

- Store precipitation and runoff, reducing water loss through evaporation.
- Enhance groundwater replenishment in regions with seasonal water scarcity.
- Improve water quality before it reaches underground reserves.

Riparian Wetlands

Wetlands found along riverbanks and streams provide critical groundwater recharge functions by stabilizing water flow and promoting infiltration. These wetlands:

- Maintain water levels in shallow aquifers connected to river systems.
- Reduce sedimentation and improve water quality before infiltration.

- Support aquatic and terrestrial ecosystems by regulating moisture availability.

Peatlands and Bogs

Although primarily known for carbon storage, peatlands also contribute to groundwater recharge by slowly releasing water stored in organic-rich soils. These wetlands:

- Retain moisture for extended periods, providing a continuous water source to surrounding landscapes.
- Release water gradually, maintaining base flow in nearby rivers and streams.
- Store water that can percolate into deeper soil layers during dry seasons.

Threats to Wetland-Groundwater Recharge Functions

Despite their importance, wetlands that contribute to groundwater recharge are increasingly under threat from human activities and environmental changes.

Wetland Drainage and Land Conversion

The drainage of wetlands for agriculture, urban development, and infrastructure projects disrupts natural infiltration processes and reduces groundwater recharge capacity.

- **Agricultural Expansion**
  Converting wetlands into cropland eliminates their ability to store and release water, leading to increased surface runoff and reduced aquifer replenishment.
- **Urbanization and Infrastructure**
  Paving over wetlands with roads, buildings, and industrial sites prevents water from infiltrating the ground, increasing the risk of groundwater depletion and urban flooding.

Groundwater Overextraction

Excessive groundwater pumping for irrigation, drinking water, and industrial use depletes aquifers faster than they can be recharged, threatening long-term water availability.

- **Irrigation Demands**
  In agricultural regions, groundwater extraction for irrigation lowers water tables, reducing the effectiveness of wetlands in replenishing underground reserves.
- **Municipal and Industrial Use**
  Cities and industries rely on groundwater for water supply, often withdrawing large amounts without adequate replenishment strategies.

Climate Change and Altered Hydrological Cycles

Changes in precipitation patterns, temperature increases, and extreme weather events affect wetland hydrology and groundwater recharge rates.

- **Drought and Reduced Rainfall**
  Extended dry periods reduce wetland water levels, limiting their ability to contribute to groundwater recharge.
- **Intense Storms and Flooding**
  While heavy rains can temporarily replenish wetlands, extreme flooding can overwhelm infiltration capacity, causing rapid runoff instead of steady groundwater recharge.

Strategies to Protect and Enhance Wetland-Groundwater Recharge

To maintain the role of wetlands in groundwater recharge, conservation and management efforts must focus on preserving natural hydrological processes and preventing further degradation.

Wetland Conservation and Restoration

- **Protecting Natural Wetlands**
  Enforcing wetland protection policies and establishing conservation areas help safeguard groundwater recharge functions.
- **Rewetting Drained Wetlands**
  Restoring former wetlands by reflooding degraded areas can re-establish their ability to store and infiltrate water.

Sustainable Land and Water Management

- **Reducing Groundwater Overuse**
  Implementing water-saving irrigation techniques and regulating groundwater extraction can prevent excessive depletion.
- **Enhancing Green Infrastructure**
  Integrating wetlands into urban planning, such as through constructed wetlands and permeable landscapes, improves groundwater recharge in developed areas.

Climate Adaptation and Resilience Strategies

- **Managing Water Flows**
  Ensuring that wetlands receive adequate water through regulated flow management helps maintain their hydrological functions.
- **Incorporating Wetlands into Water Security Planning**
  Recognizing wetlands as essential components of water resource management can improve long-term water sustainability.

Balancing Competing Water Uses

Water is a finite resource that must be managed carefully to meet the diverse needs of ecosystems, agriculture, industry, and human populations. Wetlands play a crucial role in regulating water availability, ensuring that different sectors can access water sustainably. However, increasing demand for water, driven by population growth, economic development, and climate change, has led to conflicts over its use. Balancing competing water needs while maintaining the ecological integrity of wetlands is essential for long-term water security.

The Challenge of Competing Water Uses

Wetlands support multiple sectors, each with distinct water requirements. Conflicts arise when water demand exceeds available supply, especially in regions with limited freshwater resources. The main competing water uses include:

- **Ecosystem Needs**: Wetlands require sufficient water to sustain biodiversity, maintain water quality, and regulate hydrological cycles.
- **Agriculture**: Farming accounts for approximately 70% of global freshwater withdrawals, with irrigation-dependent crops placing high demands on water resources.
- **Urban and Domestic Use**: Cities and towns require water for drinking, sanitation, and industrial processes, leading to increased competition for freshwater sources.
- **Industry and Energy Production**: Manufacturing, mining, and hydropower generation rely heavily on water, often diverting it from natural ecosystems.

Finding a balance between these competing demands is essential to prevent overuse, degradation, and long-term water shortages.

The Role of Wetlands in Balancing Water Use

Wetlands act as natural regulators of water availability, helping to mediate competing demands by storing, filtering, and gradually releasing water. Their role in sustainable water management includes:

- **Water Storage and Regulation**: Wetlands retain excess water during rainy seasons and release it slowly during dry periods, stabilizing water supply for multiple users.
- **Flood and Drought Mitigation**: By absorbing floodwaters and maintaining base flows during droughts, wetlands help ensure consistent water availability for agriculture, cities, and industry.
- **Groundwater Recharge**: Wetlands contribute to groundwater replenishment, sustaining aquifers that supply drinking water, irrigation, and industrial needs.
- **Water Filtration**: Natural wetland processes improve water quality, making it safer for human consumption and industrial use.

Ensuring that wetlands receive adequate water to continue providing these benefits is key to managing competing demands effectively.

Conflicts Between Water Users and Wetland Conservation

Despite their benefits, wetlands are often degraded or drained to meet human water needs, leading to conflicts between conservation and economic development.

Agriculture vs. Wetland Conservation

Agriculture is the largest consumer of freshwater, often competing directly with wetlands for water resources. Common conflicts include:

- **Irrigation Demands**: Diverting water for irrigation can reduce wetland water levels, threatening ecosystem health.
- **Agrochemical Pollution**: Fertilizers and pesticides from farmlands contaminate wetland waters, affecting biodiversity and water quality.
- **Land Conversion**: Wetlands are frequently drained to create farmland, reducing their ability to regulate water and support biodiversity.

Sustainable agricultural practices, such as precision irrigation, agroforestry, and buffer zones, can help balance water needs without compromising wetland integrity.

Urban Expansion and Water Supply Conflicts

Rapid urbanization increases water demand for drinking, sanitation, and industrial use, often leading to wetland degradation.

- **Water Extraction for Municipal Supply**: Cities pump large volumes of water from wetlands and nearby rivers, reducing water availability for ecosystems.
- **Infrastructure Development**: Roads, buildings, and industrial zones alter natural drainage patterns, affecting wetland hydrology.
- **Wastewater Discharge**: Urban wastewater, if untreated, degrades wetland water quality, impacting biodiversity and ecosystem services.

Integrating wetlands into urban planning through green infrastructure and sustainable drainage systems can help cities meet water needs while preserving natural ecosystems.

Industry and Energy Production Impacts

Industries and power plants depend on wetlands for cooling water, waste disposal, and raw materials, leading to conflicts with conservation efforts.

- **Hydropower Projects**: Dams and reservoirs alter natural water flows, disrupting wetland ecosystems and migratory species.
- **Mining and Industrial Pollution**: Heavy metal contamination from industrial discharge affects wetland water quality, harming aquatic life.

- **Thermal Pollution**: Power plants release heated water into wetlands, altering aquatic ecosystems and reducing oxygen levels.

Adopting cleaner production methods, enforcing environmental regulations, and investing in water-efficient technologies can reduce industrial impacts on wetlands.

Strategies for Balancing Competing Water Uses

Ensuring sustainable water management requires policies and practices that recognize the needs of all users while protecting wetland ecosystems.

Integrated Water Resource Management (IWRM)

IWRM promotes a coordinated approach to managing water resources, balancing social, economic, and environmental needs. Key principles include:

- **Cross-Sector Coordination**: Engaging stakeholders from agriculture, urban planning, industry, and conservation to develop shared water management strategies.
- **Water Allocation Planning**: Establishing fair water distribution frameworks that prioritize ecosystem health while supporting human needs.
- **Adaptive Management**: Using data and monitoring to adjust water use policies based on changing environmental conditions.

Sustainable Agricultural Practices

Reducing agricultural water consumption and pollution can help maintain wetland health while ensuring food security. Sustainable approaches include:

- **Drip and Precision Irrigation**: Using targeted irrigation techniques to reduce water waste and maintain soil moisture.
- **Wetland Buffers and Riparian Zones**: Establishing vegetated buffer zones around wetlands to filter agricultural runoff and protect water quality.
- **Crop Rotation and Agroecology**: Promoting farming methods that enhance soil fertility and reduce dependency on chemical inputs.

Urban and Industrial Water Management

Cities and industries can improve water efficiency and reduce wetland degradation through:

- **Water Recycling and Reuse**: Implementing wastewater treatment and reuse systems to reduce freshwater extraction.
- **Stormwater Management**: Using green infrastructure such as permeable pavements and retention ponds to prevent urban flooding and enhance groundwater recharge.
- **Strict Pollution Control Regulations**: Enforcing policies that limit industrial discharge and promote cleaner production technologies.

Community Involvement and Policy Integration

Public participation and strong governance are critical for achieving balanced water use while protecting wetlands.

- **Stakeholder Engagement**: Involving local communities, farmers, businesses, and conservation groups in decision-making ensures fair water allocation.
- **Environmental Protection Laws**: Strengthening legal frameworks to protect wetlands and enforce sustainable water use practices.
- **Public Awareness Campaigns**: Educating communities about the importance of wetlands in water security encourages conservation efforts.

Chapter 5: Wetlands and Sustainable Development Goals

Wetlands are essential ecosystems that contribute to achieving multiple Sustainable Development Goals (SDGs) by providing clean water, supporting biodiversity, mitigating climate change, and sustaining livelihoods. Their role in water purification, food security, disaster risk reduction, and carbon sequestration directly aligns with global efforts to promote environmental sustainability, economic growth, and social well-being.

However, wetland degradation threatens progress toward these goals. Unsustainable land use, pollution, and climate change are reducing their capacity to provide essential services. Protecting and restoring wetlands is critical for achieving SDG targets related to water security, climate resilience, and biodiversity conservation.

This chapter explores the connections between wetlands and the SDGs, highlighting their contributions to sustainable water management, climate action, economic development, and human well-being. It also examines the challenges of integrating wetlands into policy frameworks and emphasizes the need for stronger conservation efforts to support long-term sustainability.

Wetlands' Direct Role in Sustainable Development

Wetlands play a fundamental role in achieving the SDGs by directly supporting environmental health, economic growth, and social well-being. Their ability to regulate water cycles, store carbon, protect biodiversity, and sustain livelihoods makes them essential for sustainable development. However, wetlands face increasing pressures from human activities, requiring stronger conservation and management strategies to maximize their contributions to global sustainability targets.

Wetlands and Their Direct Role

Wetlands provide essential ecosystem services that directly contribute to multiple SDGs, particularly those related to water security, climate action, biodiversity conservation, and sustainable resource management.

SDG 6: Clean Water and Sanitation

Wetlands are vital for maintaining water quality, ensuring freshwater availability, and supporting sanitation efforts. Their natural filtration processes remove pollutants, sediments, and excess nutrients, improving water quality for human consumption, agriculture, and industry.

- **Water Filtration**: Wetlands absorb contaminants such as heavy metals, pesticides, and organic pollutants, making them key to clean water supply systems.
- **Groundwater Recharge**: Many wetlands contribute to groundwater replenishment, ensuring long-term access to freshwater resources.
- **Wastewater Treatment**: Constructed wetlands are increasingly used as cost-effective, sustainable wastewater treatment solutions for urban and rural communities.

By protecting and restoring wetlands, governments can improve water quality, expand sanitation access, and reduce waterborne diseases, directly contributing to SDG 6 targets.

SDG 13: Climate Action

Wetlands play a crucial role in climate change mitigation and adaptation by sequestering carbon, regulating temperatures, and reducing climate-related risks such as floods and droughts.

- **Carbon Sequestration**: Coastal wetlands, including mangroves, salt marshes, and seagrass meadows, store vast amounts of carbon, preventing greenhouse gas emissions.

- **Flood and Storm Protection**: Wetlands absorb excess water during extreme weather events, reducing flood damage in vulnerable areas.
- **Drought Resilience**: By retaining water, wetlands help sustain freshwater availability during prolonged dry periods.

Investing in wetland conservation strengthens climate resilience, aligning with SDG 13 goals to reduce climate-related disasters and enhance adaptive capacities.

SDG 14: Life Below Water

Coastal and marine wetlands, such as mangroves, coral reef-associated wetlands, and estuaries, provide critical habitat for marine biodiversity and support sustainable fisheries.

- **Breeding and Nursery Grounds**: Many fish and shellfish species depend on wetlands for reproduction, making them essential for fisheries and food security.
- **Water Quality Maintenance**: Wetlands trap pollutants before they enter marine environments, protecting ocean ecosystems.
- **Coastal Protection**: Mangroves and salt marshes stabilize shorelines, reducing erosion and safeguarding marine habitats.

Protecting wetlands contributes to SDG 14 objectives by preserving marine biodiversity, ensuring sustainable fishing practices, and preventing ocean pollution.

SDG 15: Life on Land

Inland wetlands, including freshwater marshes, floodplains, and peatlands, are home to diverse plant and animal species, playing a crucial role in terrestrial biodiversity conservation.

- **Habitat for Endangered Species**: Wetlands support a wide range of birds, mammals, amphibians, and insects, many of which are threatened by habitat destruction.
- **Soil and Water Conservation**: Wetlands prevent land degradation by stabilizing soil, preventing erosion, and maintaining hydrological balance.
- **Ecosystem Connectivity**: Many wetlands serve as migration corridors for birds and other species, ensuring genetic diversity and ecological stability.

By preventing wetland degradation and promoting ecosystem restoration, SDG 15 targets for biodiversity conservation and land restoration can be achieved.

SDG 2: Zero Hunger

Wetlands support food security through sustainable fisheries, aquaculture, and agriculture, providing livelihoods for millions of people worldwide.

- **Fish and Aquatic Resources**: Wetlands provide habitat for fish, shellfish, and aquatic plants that contribute to local diets and global food supplies.
- **Floodplain Agriculture**: Many communities rely on floodplains for nutrient-rich soils that enhance agricultural productivity.
- **Livelihood Support**: Wetlands sustain rural economies by providing resources for fishing, farming, and traditional food production.

Integrating wetland conservation into agricultural planning supports SDG 2 goals by ensuring long-term food production and resource sustainability.

SDG 11: Sustainable Cities and Communities

Urban wetlands provide essential services that enhance the livability and resilience of cities.

- **Stormwater Management**: Wetlands reduce urban flooding by absorbing excess rainfall and filtering stormwater runoff.
- **Green Spaces and Recreation**: Urban wetlands improve quality of life by providing recreational spaces, enhancing mental well-being, and promoting biodiversity in cities.
- **Air Quality and Cooling Effects**: Wetlands help regulate urban temperatures, reducing heat island effects and improving air quality.

Protecting wetlands within urban planning frameworks aligns with SDG 11 by fostering sustainable and resilient cities.

Challenges in Maximizing Wetlands' Contributions to SDGs

Despite their significant role in sustainable development, wetlands face multiple threats that hinder their ability to contribute to SDG targets.

- **Wetland Loss and Degradation**: Human activities, including land conversion, pollution, and overextraction of water resources, reduce wetland functionality.
- **Lack of Policy Integration**: Many national development plans do not fully recognize the importance of wetlands, leading to insufficient protection and investment.
- **Competing Land Uses**: Urbanization, industrialization, and agricultural expansion often prioritize short-term economic gains over long-term wetland conservation.

Addressing these challenges requires stronger environmental policies, cross-sector collaboration, and increased public awareness of wetlands' value in achieving sustainable development.

Strategies for Strengthening Wetlands' Role in SDGs

To maximize wetlands' contributions to SDGs, targeted conservation and management strategies must be implemented at local, national, and global levels.

- **Integrating Wetlands into National Policies**: Governments should include wetland protection in water management, climate adaptation, and land-use planning policies.
- **Expanding Wetland Restoration Efforts**: Large-scale restoration initiatives can enhance wetlands' ability to sequester carbon, filter water, and support biodiversity.
- **Strengthening Community Involvement**: Engaging local communities in wetland conservation ensures sustainable resource use and promotes traditional knowledge.
- **Increasing Financial Investments**: Governments, businesses, and international organizations should invest in wetland conservation projects to enhance their ecosystem services.

Wetlands' Indirect Role in Sustainable Development

Wetlands provide essential ecosystem services that contribute to the SDGs not only directly but also through indirect pathways that support environmental stability, economic resilience, and social well-being. While their primary role in water security, climate mitigation, and biodiversity conservation is well recognized, wetlands also influence sustainable development by enhancing economic opportunities, cultural heritage, disaster resilience, and human health. However, wetland degradation and mismanagement threaten these indirect benefits, underscoring the need for stronger conservation efforts.

Wetlands' Indirect Role

Wetlands support multiple SDGs through their influence on economic activities, climate regulation, disaster resilience, and community well-being. These indirect contributions are often overlooked but play a crucial role in achieving global sustainability targets.

SDG 1: No Poverty – Supporting Livelihoods and Economic Stability

Wetlands indirectly contribute to poverty alleviation by sustaining economic activities that rely on natural resources. Many communities, particularly in developing regions, depend on wetlands for income, employment, and food security.

- **Sustainable Livelihoods**: Wetlands support diverse economic activities, including fishing, ecotourism, and wetland-based crafts, providing jobs for millions of people.
- **Economic Buffering Against Environmental Shocks**: Healthy wetlands reduce the financial impact of disasters such as floods and droughts, which often push vulnerable communities into poverty.
- **Provision of Natural Resources**: Wetlands provide materials like reeds, timber, and medicinal plants that support small-scale industries and traditional practices.

By maintaining wetland health, communities can strengthen their economic resilience and reduce poverty levels.

SDG 3: Good Health and Well-Being – Enhancing Public Health

Wetlands indirectly support public health by improving air and water quality, reducing disease risks, and providing spaces for recreation and mental well-being.

- **Natural Water Purification**: Wetlands filter pollutants and pathogens from water sources, reducing the incidence of waterborne diseases such as cholera and dysentery.

- **Regulation of Air Quality**: Wetland vegetation absorbs pollutants and improves air quality, benefiting respiratory health, particularly in urban areas.
- **Mental and Physical Well-Being**: Wetlands provide green spaces for recreation, reducing stress and promoting physical activity. Many studies show that access to nature improves mental health and reduces anxiety.

The degradation of wetlands can lead to increased disease risks, poorer air quality, and a decline in overall public health.

SDG 8: Decent Work and Economic Growth – Promoting Sustainable Economies

Wetlands indirectly contribute to sustainable economic growth by supporting industries such as tourism, fisheries, and sustainable agriculture.

- **Ecotourism and Nature-Based Recreation**: Many wetlands attract tourists for birdwatching, kayaking, and wildlife safaris, generating revenue for local economies.
- **Sustainable Fisheries and Aquaculture**: Wetlands provide breeding grounds for commercially valuable fish species, supporting food industries and international trade.
- **Agricultural Productivity**: Floodplain wetlands replenish soils with nutrients, reducing the need for synthetic fertilizers and promoting sustainable farming practices.

Sustaining wetland ecosystems enhances economic stability and promotes green job opportunities in conservation, tourism, and resource management.

SDG 9: Industry, Innovation, and Infrastructure – Enhancing Sustainable Development Strategies

Wetlands influence infrastructure planning and innovation by providing natural solutions for water management, flood control, and climate adaptation.

- **Natural Flood Management**: Wetlands reduce the need for costly flood defense infrastructure by acting as natural buffers against storm surges and extreme rainfall.
- **Sustainable Urban Planning**: Wetlands contribute to NbS in cities, such as green roofs, wetland parks, and stormwater management systems.
- **Innovation in Wastewater Treatment**: Constructed wetlands are increasingly used in industrial and municipal wastewater treatment, offering cost-effective and environmentally friendly solutions.

Incorporating wetlands into infrastructure planning reduces development costs while enhancing sustainability.

SDG 12: Responsible Consumption and Production – Supporting Sustainable Resource Use

Wetlands indirectly promote sustainable consumption by providing renewable resources, reducing waste, and supporting circular economy initiatives.

- **Sustainable Harvesting of Wetland Products**: Wetlands provide food, fiber, and medicinal plants that can be sustainably harvested for local and global markets.
- **Reduction of Industrial Pollution**: Wetlands filter waste from agricultural and industrial activities, preventing pollutants from entering water systems.
- **Enhancing Circular Economy Practices**: Many wetland plants, such as reeds and water hyacinths, are used in biodegradable packaging, paper production, and bioenergy.

Sustaining wetlands aligns with circular economy principles by promoting resource efficiency and reducing environmental impacts.

SDG 16: Peace, Justice, and Strong Institutions – Supporting Governance and Conflict Prevention

Water scarcity and environmental degradation are key drivers of conflicts, particularly in regions where wetlands provide essential water resources. Effective wetland management contributes to peace and governance by ensuring fair water distribution and protecting natural rights.

- **Preventing Water Conflicts**: Healthy wetlands support equitable water availability, reducing competition and tensions over water access.
- **Strengthening Environmental Governance**: International agreements, such as the Ramsar Convention, promote transboundary wetland conservation, fostering cooperation between nations.
- **Enhancing Community Participation**: Wetland conservation projects that involve local communities improve governance structures and empower marginalized groups.

Integrating wetlands into global and national governance frameworks enhances stability, equity, and environmental justice.

Threats to Wetlands' Indirect Contributions

Despite their importance, wetlands face increasing pressures that threaten their indirect contributions to sustainable development.

- **Wetland Conversion and Habitat Loss**: Draining wetlands for agriculture, infrastructure, and urban expansion reduces their ability to provide ecosystem services.
- **Pollution and Water Contamination**: Industrial waste, plastic pollution, and agricultural runoff degrade wetland water quality, impacting health, fisheries, and food security.
- **Climate Change Impacts**: Rising temperatures, changing precipitation patterns, and extreme weather events threaten wetland ecosystems and their role in disaster mitigation.

- **Lack of Policy Integration**: Many national and international policies do not fully recognize the role of wetlands in achieving SDGs, leading to weak conservation efforts.

Addressing these challenges requires stronger policy interventions, investment in wetland restoration, and greater public awareness of their benefits.

Enhancing Wetlands' Indirect Contributions

To maximize wetlands' indirect contributions to the SDGs, sustainable management strategies must be prioritized.

- **Integrating Wetlands into National Development Plans**: Governments should incorporate wetland conservation into policies related to climate adaptation, water management, and economic planning.
- **Promoting Sustainable Economic Activities**: Encouraging ecotourism, sustainable fisheries, and wetland-based industries can generate economic benefits without degrading ecosystems.
- **Strengthening Wetland Protection Laws**: Enforcing regulations that prevent wetland destruction and pollution will help maintain their ecosystem functions.
- **Investing in Wetland Education and Awareness**: Public engagement campaigns and community-based conservation initiatives can enhance wetland stewardship.

Cross-Sectoral Benefits

Wetlands provide cross-sectoral benefits that extend beyond environmental conservation, influencing industries, economies, public health, and social development. Their ability to regulate water, store carbon, sustain biodiversity, and support livelihoods makes them integral to multiple sectors, from agriculture and tourism to infrastructure and disaster management. The interconnected nature of wetlands means that their protection and restoration can generate widespread advantages across various industries and policy areas. However, wetlands remain undervalued in many decision-making processes, leading to degradation and lost opportunities for sustainable development. Recognizing their cross-sectoral benefits is essential for integrating wetlands into broader economic and policy frameworks.

Wetlands and Their Influence Across Sectors

Wetlands support multiple industries and services, enhancing economic resilience, climate adaptation, and sustainable development. Their benefits are felt across key sectors, including water management, agriculture, fisheries, tourism, infrastructure, and disaster risk reduction.

Water Management and Security

Wetlands play a crucial role in water regulation, benefiting municipal water supply systems, hydropower, and sanitation services.

- **Water Storage and Regulation**: Wetlands capture excess rainwater, recharge groundwater, and release stored water gradually, stabilizing water availability during droughts and preventing floods.
- **Water Quality Improvement**: Acting as natural filtration systems, wetlands remove pollutants, sediments, and excess nutrients, reducing the need for expensive water treatment infrastructure.
- **Municipal and Industrial Water Supply**: Many cities and industries rely on wetlands to maintain water sources. Degraded wetlands can lead to increased water treatment costs and supply shortages.

Investing in wetland conservation enhances long-term water security, reducing costs for municipalities, industries, and households.

Agriculture and Food Security

Wetlands provide essential ecosystem services that support agriculture and food production, benefiting farmers and rural communities.

- **Nutrient-Rich Soils**: Floodplain wetlands replenish soil fertility through sediment deposition, reducing the need for chemical fertilizers.
- **Water Availability for Irrigation**: Many agricultural areas rely on wetlands for consistent water supply, particularly in arid and semi-arid regions.
- **Pollination and Pest Control**: Wetland habitats support pollinators, such as bees and butterflies, while also housing natural predators that control agricultural pests.

However, agriculture is also a major driver of wetland loss, with drainage and water extraction reducing wetlands' ability to support sustainable farming. Integrating wetland conservation into agricultural policies can help balance food production with ecosystem protection.

Fisheries and Aquaculture

Wetlands are essential for fisheries and aquaculture, supporting both wild-caught and farmed fish industries.

- **Breeding and Nursery Grounds**: Many commercially important fish and shellfish species depend on wetlands for spawning and early development.
- **Sustaining Inland and Coastal Fisheries**: Rivers, lakes, estuaries, and mangrove forests provide food and livelihoods for millions of people involved in fishing.
- **Water Quality Maintenance for Aquaculture**: Wetlands help maintain water quality in fish farming areas by filtering pollutants and stabilizing aquatic ecosystems.

Unsustainable fishing practices, pollution, and wetland destruction threaten fisheries-dependent economies. Protecting wetlands ensures the sustainability of fish stocks and secures livelihoods in fishing communities.

Tourism and Recreation

Wetlands attract millions of visitors annually, supporting ecotourism, nature-based recreation, and cultural tourism.

- **Wildlife Viewing and Birdwatching**: Many wetlands are home to migratory birds and rare wildlife, drawing ecotourists and supporting local tourism economies.
- **Outdoor Recreation**: Activities such as kayaking, fishing, hiking, and photography contribute to the outdoor recreation industry.
- **Cultural and Historical Significance**: Many indigenous and local communities have deep cultural ties to wetlands, which hold historical, spiritual, and recreational value.

Sustainable wetland tourism can generate economic benefits while funding conservation efforts. However, poorly managed tourism can lead to habitat degradation and pollution, requiring careful planning and regulation.

Infrastructure and Urban Development

Urban wetlands provide multiple benefits for cities, improving resilience to climate change and reducing infrastructure costs.

- **Flood Control**: Wetlands absorb and store excess stormwater, reducing the impact of floods on urban areas and infrastructure.
- **Stormwater Management**: Wetlands help filter and regulate urban runoff, reducing pressure on drainage systems and lowering water treatment costs.

- **Heat Island Reduction**: Wetland vegetation cools surrounding areas, mitigating the effects of urban heat islands and reducing energy consumption for cooling.

Incorporating wetlands into urban planning enhances climate resilience, lowers disaster-related damages, and improves overall quality of life for city residents.

Disaster Risk Reduction

Wetlands serve as natural buffers against extreme weather events, reducing disaster risks and protecting vulnerable communities.

- **Storm Surge and Coastal Protection**: Mangroves and salt marshes reduce the impact of hurricanes and storm surges by dissipating wave energy.
- **Drought Resilience**: Wetlands maintain groundwater levels, ensuring water availability during prolonged dry spells.
- **Landslide and Erosion Control**: Wetland vegetation stabilizes soil, preventing erosion and reducing the likelihood of landslides in hilly and mountainous regions.

Investing in wetland conservation as part of disaster risk management strategies can significantly reduce economic losses and improve community resilience.

Barriers to Maximizing Wetlands' Cross-Sectoral Benefits

Despite their contributions to multiple sectors, wetlands are often undervalued and underprotected due to policy, economic, and governance challenges.

- **Competing Land-Use Demands**: Agriculture, urban expansion, and industrial development frequently take priority over wetland conservation.

- **Fragmented Policies**: Many water, agriculture, and urban planning policies fail to integrate wetland conservation, leading to conflicting land-use decisions.
- **Short-Term Economic Pressures**: Industries often prioritize immediate economic gains over long-term sustainability, resulting in wetland degradation.
- **Limited Public Awareness**: Many policymakers and businesses are unaware of the full economic and environmental benefits wetlands provide, leading to insufficient investment in their protection.

Addressing these barriers requires cross-sector collaboration, policy integration, and stronger incentives for sustainable wetland management.

Enhancing Wetlands' Cross-Sectoral Contributions

To maximize the cross-sectoral benefits of wetlands, governments, industries, and communities must adopt integrated approaches that balance conservation with economic development.

- **Incorporating Wetlands into National Policies**: Wetlands should be included in water management, climate adaptation, and land-use planning strategies.
- **Investing in NbS**: Governments and businesses can prioritize wetlands as cost-effective solutions for flood control, carbon sequestration, and water filtration.
- **Developing Sustainable Financing Mechanisms**: Incentives such as wetland conservation credits, carbon markets, and ecotourism revenue-sharing can support wetland protection.
- **Strengthening Public Awareness and Education**: Increasing knowledge about wetland benefits can drive community engagement and support for conservation initiatives.

Chapter 6: Restoration and Conservation of Wetlands

Wetlands provide essential ecological services, yet they are rapidly disappearing due to land conversion, pollution, and climate change. Their loss threatens biodiversity, water security, and climate resilience.

Restoring and conserving wetlands is crucial for reversing degradation and sustaining their benefits. This requires protecting existing wetlands, rehabilitating damaged areas, and integrating conservation into broader policies.

This chapter explores key strategies for wetland restoration and conservation, addressing challenges and highlighting the role of governments, businesses, and communities in ensuring their long-term sustainability.

Wetland Restoration Strategies

Restoring wetlands is essential for reversing ecosystem degradation, improving water quality, enhancing biodiversity, and strengthening climate resilience. Wetland restoration involves re-establishing natural hydrological processes, reintroducing native vegetation, and removing human-induced disturbances. Given the rapid decline of wetlands worldwide due to land conversion, pollution, and climate change, effective restoration strategies are critical for maintaining the long-term ecological and economic benefits these ecosystems provide.

Wetland restoration efforts focus on restoring natural functions and processes that have been disrupted by human activities. The most effective strategies depend on the type of wetland, the degree of degradation, and the surrounding land-use context.

Hydrological Restoration

Re-establishing the natural water flow is the foundation of wetland restoration, as altered hydrology is one of the primary causes of wetland degradation.

- **Rewetting Drained Wetlands**: Many wetlands have been drained for agriculture, urban expansion, and infrastructure development. Restoring these wetlands involves blocking drainage channels, removing pumps, and allowing water levels to return to their natural state.
- **Restoring Floodplain Connectivity**: Floodplain wetlands rely on seasonal flooding to sustain their ecological functions. Removing levees and reconnecting wetlands to rivers can restore natural flood cycles and improve water retention.
- **Managing Water Flows**: In some cases, controlled water releases from reservoirs and dams can mimic natural hydrological patterns, ensuring wetlands receive adequate water.

Hydrological restoration improves groundwater recharge, enhances habitat quality, and increases the resilience of wetlands to climate-related changes.

Reintroducing Native Vegetation

Restoring plant communities is essential for stabilizing wetland ecosystems, improving water quality, and providing habitat for wildlife.

- **Planting Native Wetland Species**: Reintroducing native reeds, sedges, mangroves, and other wetland vegetation helps restore ecosystem functions and prevent erosion.
- **Removing Invasive Species**: Non-native plants, such as water hyacinth and common reed (Phragmites australis), can outcompete native vegetation and alter wetland dynamics. Removing invasive species allows native plants to thrive.

- **Encouraging Natural Regeneration**: In some cases, simply restoring natural water conditions allows native vegetation to recolonize the area without direct human intervention.

Healthy wetland vegetation improves carbon sequestration, enhances biodiversity, and stabilizes shorelines against erosion.

Improving Water Quality and Reducing Pollution

Water pollution from agriculture, industry, and urban runoff significantly degrades wetland ecosystems. Restoration efforts focus on reducing contaminant inflows and improving water quality.

- **Establishing Buffer Zones**: Creating vegetated buffer strips around wetlands helps filter pollutants before they enter wetland systems.
- **Reducing Agricultural Runoff**: Implementing sustainable farming practices, such as precision fertilization and wetland-friendly irrigation, minimizes nutrient and pesticide pollution.
- **Restoring Natural Filtration Functions**: Constructed wetlands and biofiltration systems can be integrated into landscapes to improve water quality in degraded wetlands.

Improving water quality enhances wetland ecosystem health and supports the recovery of aquatic life.

Reintroducing Wildlife and Restoring Habitat Connectivity

Many wetland-dependent species have declined due to habitat fragmentation and degradation. Restoration efforts focus on improving habitat conditions and supporting species recovery.

- **Providing Wildlife Corridors**: Connecting fragmented wetlands through ecological corridors allows species to migrate, breed, and access necessary resources.

- **Reintroducing Key Species**: In some cases, species such as beavers, amphibians, and fish need to be reintroduced to restore ecological functions. Beavers, for example, naturally enhance wetlands by building dams that create water-retaining habitats.
- **Creating Artificial Nesting and Breeding Sites**: For species with declining populations, artificial structures like floating platforms or fish spawning beds can support recovery efforts.

Restoring biodiversity improves wetland resilience and strengthens ecological interactions.

Managing Climate Adaptation and Resilience

Wetland restoration must account for climate change impacts, including rising temperatures, altered rainfall patterns, and sea-level rise.

- **Enhancing Coastal Wetlands**: Mangrove restoration and salt marsh rehabilitation protect shorelines from erosion and storm surges.
- **Restoring Peatlands for Carbon Sequestration**: Peatlands store large amounts of carbon, and their restoration helps mitigate climate change.
- **Using Adaptive Management Strategies**: Monitoring environmental conditions and adjusting restoration techniques as needed ensures long-term success.

Incorporating climate resilience into restoration planning helps wetlands adapt to changing conditions and continue providing essential services.

Challenges in Wetland Restoration

Despite the benefits of wetland restoration, several challenges must be addressed to ensure long-term success.

- **Land-Use Conflicts**: Competing demands for agriculture, urban expansion, and industry often take priority over wetland conservation efforts.
- **High Restoration Costs**: Large-scale restoration projects require significant financial investment, making funding a barrier to implementation.
- **Lack of Public Awareness**: Many wetland benefits are undervalued, leading to insufficient support for restoration initiatives.
- **Monitoring and Maintenance Needs**: Wetland restoration requires long-term monitoring to assess progress and address emerging threats.

Overcoming these challenges requires integrated policies, financial incentives, and community engagement.

Strategies for Successful Wetland Restoration

To enhance the effectiveness of wetland restoration projects, governments, organizations, and communities can adopt the following approaches:

- **Integrating Wetlands into Land-Use Planning**: Ensuring that wetland conservation is included in regional and national planning frameworks can prevent further degradation.
- **Expanding Funding and Incentives**: Governments and businesses can invest in wetland restoration through carbon credits, conservation grants, and public-private partnerships.
- **Promoting Community Involvement**: Engaging local communities in restoration projects fosters stewardship and ensures long-term sustainability.
- **Enhancing Policy and Legal Protections**: Strengthening wetland protection laws and enforcing conservation regulations prevent further degradation and support restoration efforts.

Wetland Protection Policies

Wetlands are among the most valuable ecosystems on Earth, providing essential services such as water filtration, flood control, biodiversity support, and climate regulation. However, they face increasing threats from urban expansion, agriculture, pollution, and climate change. To safeguard these ecosystems, wetland protection policies play a crucial role in regulating land use, promoting conservation, and ensuring sustainable management. Effective policies at local, national, and international levels help maintain wetland functions while balancing economic development and environmental sustainability.

The Importance of Wetland Protection Policies

Wetland protection policies are necessary to:

- **Prevent Wetland Loss**: Policies establish legal protections to prevent drainage, land conversion, and destruction.
- **Regulate Land and Water Use**: Laws and guidelines ensure that industries, agriculture, and urban development minimize their impact on wetlands.
- **Support Climate Resilience**: Wetland conservation policies help combat climate change by preserving carbon sinks such as peatlands and mangroves.
- **Maintain Water Quality**: Regulations control pollution sources, such as agricultural runoff and industrial waste, to protect wetland ecosystems.
- **Enhance Biodiversity Conservation**: Many wetland species rely on protected areas for habitat, breeding, and migration.

Without strong policies, wetland degradation accelerates, leading to loss of ecosystem services and increased vulnerability to environmental changes.

Key Components of Effective Wetland Protection Policies

Successful wetland policies incorporate legal frameworks, conservation incentives, monitoring programs, and enforcement mechanisms.

Legal Frameworks and Designation of Protected Areas

Governments establish laws and regulations to designate wetlands as protected areas, ensuring their long-term conservation.

- **Wetland Reserves and National Parks**: Many countries create wetland reserves to prohibit harmful activities such as draining, logging, and pollution.
- **International Treaties and Agreements**: Global frameworks, such as the Ramsar Convention, support the protection of internationally significant wetlands.
- **Local Zoning Regulations**: Municipal governments can enforce zoning laws that restrict construction, farming, and industrial activities in wetland areas.

Protected status ensures that wetlands remain intact and function as natural water regulators, biodiversity hotspots, and climate buffers.

Incentives for Wetland Conservation

Economic and financial incentives encourage landowners, businesses, and communities to engage in wetland protection.

- **Payment for Ecosystem Services (PES)**: Programs compensate landowners for maintaining wetlands instead of converting them for agriculture or development.
- **Tax Benefits and Subsidies**: Governments provide tax deductions or financial incentives for wetland conservation efforts.
- **Sustainable Land-Use Practices**: Encouraging eco-friendly farming, wetland-compatible grazing, and sustainable forestry reduces wetland degradation.

By providing economic incentives, wetland policies align conservation goals with financial interests, making protection efforts more sustainable.

Pollution Control and Water Quality Regulations

Strict pollution control measures protect wetlands from contamination caused by industrial waste, agricultural runoff, and urban stormwater.

- **Regulating Industrial Discharge**: Policies set limits on chemical, heavy metal, and wastewater discharge into wetland ecosystems.
- **Managing Agricultural Runoff**: Buffer zones, wetland-friendly irrigation, and nutrient management plans help prevent excessive fertilizer and pesticide pollution.
- **Stormwater Management Strategies**: Cities implement wetland-based filtration systems to improve water quality in urban environments.

These regulations help maintain wetland health, ensuring that they continue to filter water, support biodiversity, and regulate nutrient cycles.

Wetland Monitoring and Assessment Programs

Regular monitoring ensures that wetland policies remain effective and adaptable to environmental changes.

- **Remote Sensing and Satellite Mapping**: Technology helps track wetland loss, land-use changes, and habitat degradation over time.
- **Water Quality Testing**: Government agencies conduct routine assessments of pollutants, nutrient levels, and hydrological conditions.

- **Biodiversity Surveys**: Conservation organizations monitor wetland-dependent species to assess ecosystem health and detect threats.

Data collected from monitoring programs inform policy adjustments and improve wetland management strategies.

Community Involvement and Stakeholder Engagement

Local communities play a vital role in wetland protection, and policies that include public participation lead to more effective conservation outcomes.

- **Community-Led Conservation Projects**: Many governments support grassroots efforts to restore and maintain local wetlands.
- **Educational Programs and Awareness Campaigns**: Public outreach initiatives highlight the importance of wetlands and encourage sustainable behaviors.
- **Collaborations Between Government, NGOs, and Private Sector**: Multi-stakeholder partnerships enhance wetland protection efforts through funding, research, and policy development.

Engaging local communities fosters a sense of ownership and responsibility, leading to stronger wetland stewardship.

Challenges in Implementing Wetland Protection Policies

Despite the importance of wetland protection, various challenges hinder policy implementation and enforcement.

- **Conflicting Land-Use Priorities**: Economic development, agriculture, and infrastructure projects often compete with conservation goals.

- **Weak Enforcement Mechanisms**: Many policies lack effective enforcement, allowing illegal land conversion and pollution to continue.
- **Insufficient Funding**: Conservation programs require long-term financial investment, which can be difficult to secure.
- **Lack of Coordination Among Agencies**: Overlapping jurisdictions and fragmented policies weaken wetland protection efforts.

Addressing these challenges requires stronger governance, increased funding, and better integration of wetland conservation into broader policy frameworks.

Future Directions for Strengthening Wetland Protection Policies

To improve the effectiveness of wetland protection, policymakers and conservation organizations must adopt integrated approaches that balance environmental sustainability with economic and social needs.

- **Mainstreaming Wetlands into National and Regional Planning**: Wetlands should be considered in water management, climate adaptation, and land-use policies.
- **Expanding International Cooperation**: Strengthening global agreements and cross-border conservation efforts can enhance wetland protection.
- **Investing in NbS**: Incorporating wetlands into flood management, wastewater treatment, and coastal protection strategies enhances their role in sustainable development.
- **Promoting Private Sector Engagement**: Encouraging businesses to invest in wetland conservation through corporate sustainability initiatives can increase funding and support.

Community Involvement

Wetlands provide essential ecological, economic, and social benefits, yet their conservation and restoration efforts often depend on community involvement. Local communities play a crucial role in protecting wetlands by contributing traditional knowledge, participating in conservation programs, and supporting sustainable resource management. Engaging communities ensures long-term success by fostering stewardship, enhancing public awareness, and integrating local perspectives into wetland policies. However, effective community participation requires collaboration between governments, non-governmental organizations (NGOs), businesses, and individuals to create inclusive and sustainable conservation initiatives.

The Importance of Community Involvement in Wetland Conservation

Community involvement is essential for the protection and sustainable management of wetlands because:

- **Local Knowledge Enhances Conservation Efforts**: Indigenous and rural communities often have deep ecological knowledge about wetlands, including water cycles, plant species, and wildlife behavior. Incorporating this knowledge into management strategies improves conservation effectiveness.
- **Stronger Community Ownership Encourages Long-Term Stewardship**: When communities are actively engaged, they are more likely to take responsibility for protecting and restoring wetlands.
- **Sustainable Livelihoods Reduce Environmental Pressures**: Wetlands provide food, water, and income for many communities. Sustainable management practices help balance environmental conservation with economic needs.
- **Public Awareness and Education Drive Policy Support**: Well-informed communities advocate for stronger wetland protections and support local conservation initiatives.

Engaging communities in wetland conservation ensures that protection efforts align with social, economic, and cultural priorities, leading to more resilient and effective outcomes.

Ways Communities Contribute to Wetland Protection

Communities contribute to wetland conservation through direct participation in restoration efforts, advocacy, sustainable resource management, and education initiatives.

Community-Led Conservation and Restoration Projects

Local initiatives play a vital role in maintaining and restoring wetlands. Many successful wetland restoration projects are community-driven and supported by conservation organizations and government programs.

- **Habitat Restoration**: Communities engage in planting native vegetation, removing invasive species, and restoring degraded wetland areas.
- **Water Management**: Farmers and landowners adopt wetland-friendly irrigation and drainage practices to maintain water levels and reduce pollution.
- **Monitoring and Data Collection**: Citizen science programs involve community members in tracking biodiversity, water quality, and climate-related changes in wetlands.

Restoration projects that involve local participation are more likely to succeed, as they align with community interests and benefit from on-the-ground knowledge.

Sustainable Livelihoods and Wetland-Based Economies

Many communities depend on wetlands for fishing, agriculture, tourism, and other economic activities. Sustainable livelihood programs help protect wetlands while ensuring economic stability.

- **Sustainable Fishing and Aquaculture**: Educating fishers about responsible fishing practices and implementing quotas help prevent overexploitation of wetland fisheries.
- **Eco-Friendly Agriculture**: Farmers adopt wetland-compatible techniques, such as agroforestry and organic farming, to reduce environmental impacts.
- **Ecotourism and Nature-Based Enterprises**: Community-run ecotourism businesses, such as guided wetland tours, birdwatching excursions, and cultural experiences, generate income while promoting conservation awareness.

Supporting wetland-friendly economic activities reduces pressure on natural resources and provides financial incentives for conservation.

Community Advocacy and Policy Engagement

Public participation in decision-making processes strengthens wetland protection policies and ensures that conservation measures reflect community priorities.

- **Local Environmental Groups and Citizen Action**: Many communities form conservation groups to advocate for wetland protection, influence policy decisions, and organize awareness campaigns.
- **Stakeholder Engagement in Policy Development**: Governments and conservation organizations involve local communities in wetland management planning to ensure policies consider local needs and knowledge.
- **Legal Rights and Land Tenure Protection**: Indigenous and rural communities advocate for legal recognition of their rights to manage and protect wetland areas, ensuring long-term conservation.

Community advocacy fosters stronger policies, encourages accountability, and builds social support for wetland conservation initiatives.

Education, Awareness, and Cultural Significance

Educational programs help build awareness about the value of wetlands and encourage future generations to participate in conservation efforts.

- **School and Youth Programs**: Environmental education initiatives introduce students to wetland ecology, wildlife, and sustainable practices.
- **Public Awareness Campaigns**: Workshops, media campaigns, and community events inform the public about wetland conservation challenges and solutions.
- **Cultural and Spiritual Connections**: Many communities view wetlands as sacred or culturally significant landscapes. Incorporating traditional beliefs into conservation strategies enhances local engagement.

By fostering environmental awareness, education initiatives create a foundation for long-term wetland stewardship.

Challenges to Community Involvement in Wetland Conservation

Despite the benefits of community engagement, several challenges limit participation in wetland conservation efforts.

- **Lack of Resources and Funding**: Many local conservation projects struggle with limited financial support and technical expertise.
- **Competing Economic Priorities**: Short-term economic pressures, such as agriculture and urban development, often take precedence over wetland conservation.
- **Limited Legal and Institutional Support**: Weak policies or lack of legal recognition for community-led conservation efforts hinder effective wetland protection.
- **Conflicts Between Stakeholders**: Disagreements between different water users, such as farmers, fishers, and industry, can create barriers to sustainable wetland management.

Addressing these challenges requires stronger partnerships, financial investments, and legal frameworks that empower local communities.

Strategies to Enhance Community Engagement in Wetland Protection

To maximize the impact of community involvement, conservation programs should focus on inclusive participation, financial incentives, and policy integration.

- **Strengthening Community-Based Organizations**: Supporting local conservation groups with funding, training, and resources helps scale up wetland restoration efforts.
- **Providing Financial Incentives for Conservation**: Payment for ecosystem services, eco-certifications, and community tourism initiatives encourage sustainable wetland management.
- **Building Multi-Stakeholder Partnerships**: Governments, NGOs, businesses, and community groups should collaborate to align conservation goals with economic and social development.
- **Ensuring Policy Support and Legal Recognition**: Recognizing indigenous and local community land rights helps protect wetlands from unsustainable exploitation.

By integrating community-led conservation into national and international wetland policies, governments can enhance long-term protection efforts.

Chapter 7: Wetlands in Urban Planning and Design

As cities expand, wetlands are increasingly threatened by urban development, yet they offer valuable solutions for sustainable urban planning and design. Wetlands regulate water flow, reduce flooding, improve air quality, and provide green spaces that enhance urban resilience and livability. Integrating wetlands into city planning can help mitigate climate risks, support biodiversity, and improve residents' well-being.

However, many urban wetlands have been drained or degraded due to land reclamation, pollution, and infrastructure expansion. To maximize their benefits, cities must adopt strategies that protect existing wetlands, restore degraded areas, and incorporate wetland-based solutions into urban design.

This chapter explores the role of wetlands in urban planning, highlighting their contributions to flood control, water management, and sustainable city design. It also examines challenges and strategies for integrating wetlands into modern urban landscapes to create healthier, more resilient cities.

Integrating Wetlands into Urban Landscapes

As cities expand, the integration of wetlands into urban landscapes offers a sustainable approach to managing environmental challenges while enhancing urban resilience. Wetlands provide multiple benefits, including flood control, water filtration, temperature regulation, and biodiversity conservation. However, urban development often leads to wetland degradation through land reclamation, pollution, and altered hydrology. Effective integration of wetlands into city planning requires policies and design approaches that balance urban growth with ecological sustainability.

The Role of Wetlands in Urban Environments

Urban wetlands contribute to the health, resilience, and functionality of cities in several ways:

- **Flood Control and Stormwater Management**: Wetlands absorb excess rainwater, reducing the risk of urban flooding and lowering pressure on drainage systems.
- **Water Filtration and Quality Improvement**: Wetlands naturally remove pollutants, sediments, and nutrients from runoff before they reach larger water bodies.
- **Temperature Regulation**: Wetland vegetation cools urban areas by reducing the heat island effect, improving overall climate comfort.
- **Biodiversity and Habitat Protection**: Wetlands support a variety of plant and animal species, creating green spaces that enhance urban ecology.
- **Recreational and Aesthetic Benefits**: Integrated wetlands improve urban landscapes by providing parks, walkways, and wildlife observation areas, enhancing residents' quality of life.

By incorporating wetlands into city planning, urban areas can become more resilient to climate change and environmental pressures.

Strategies for Integrating Wetlands into Urban Planning

Successful integration of wetlands into urban landscapes requires strategic planning, innovative design, and regulatory support. Key strategies include the following.

Protecting and Restoring Existing Urban Wetlands

Preserving natural wetlands within urban environments ensures they continue providing ecosystem services.

- **Enforcing Wetland Protection Laws**: Stronger regulations prevent wetland destruction due to construction, pollution, and encroachment.
- **Restoring Degraded Wetlands**: Revitalizing damaged wetlands through replanting native vegetation and restoring water flows enhances their ecological function.
- **Establishing Wetland Buffers**: Creating buffer zones around wetlands reduces human impact and protects them from urban runoff and pollutants.

Restoring and safeguarding wetlands strengthens their role in urban sustainability.

Designing Constructed Wetlands for Stormwater Management

Artificial or constructed wetlands are engineered ecosystems designed to replicate natural wetland functions, particularly for water treatment and flood mitigation.

- **Stormwater Retention Ponds**: These wetlands collect and slow stormwater runoff, reducing flooding and improving water filtration.
- **Green Infrastructure Integration**: Wetlands can be incorporated into green roofs, bioswales, and urban drainage systems to manage rainwater efficiently.
- **Multi-Functional Spaces**: Combining wetland features with recreational parks or urban green spaces maximizes their utility while maintaining ecological integrity.

Constructed wetlands provide a practical solution for cities facing increasing water management challenges.

Incorporating Wetlands into Urban Development Plans

Urban planning policies should prioritize wetland conservation and integration in land-use decisions.

- **Zoning Regulations for Wetland Conservation**: Implementing zoning laws that designate wetlands as protected areas ensures long-term sustainability.
- **Low-Impact Development (LID) Approaches**: Urban projects can use permeable surfaces, rain gardens, and natural drainage systems to reduce wetland degradation.
- **Sustainable Urban Expansion**: Encouraging development patterns that incorporate wetlands into city design prevents unnecessary habitat loss.

By embedding wetlands into planning frameworks, cities can enhance resilience while promoting sustainable growth.

Enhancing Public Engagement and Community Stewardship

Public participation in wetland conservation fosters long-term protection and responsible urban planning.

- **Community-Led Wetland Restoration**: Engaging local communities in restoration projects ensures wetlands remain valued public assets.
- **Educational Programs and Awareness Campaigns**: Raising awareness about urban wetland benefits encourages public support for conservation efforts.
- **Accessible Wetland Parks**: Developing urban wetland parks with boardwalks, trails, and educational signage enhances appreciation for these ecosystems.

Involving communities in wetland integration helps secure long-term conservation and sustainability.

Challenges to Urban Wetland Integration

Despite their benefits, integrating wetlands into urban landscapes presents several challenges:

- **Competing Land-Use Priorities**: Urban expansion and infrastructure projects often take precedence over wetland conservation.
- **Pollution and Contamination**: Industrial runoff, untreated sewage, and chemical waste degrade urban wetlands, reducing their effectiveness.
- **Fragmentation and Habitat Loss**: Urban sprawl isolates wetland ecosystems, limiting species movement and reducing biodiversity.
- **Regulatory and Policy Gaps**: Inconsistent or weak policies hinder wetland protection and restoration efforts.

Addressing these challenges requires stronger regulatory frameworks, improved planning, and sustainable urban development approaches.

Future Directions for Urban Wetland Integration

To maximize the role of wetlands in urban resilience, cities must adopt forward-thinking policies and innovative planning approaches.

- **Strengthening Wetland Protection Policies**: Governments should enforce conservation laws and integrate wetlands into climate adaptation strategies.
- **Investing in Green Infrastructure**: Expanding urban wetland projects, such as rain gardens and floodplain restorations, enhances water management.
- **Promoting Public-Private Partnerships**: Collaboration between municipalities, businesses, and conservation groups can fund wetland integration projects.
- **Developing Climate-Resilient Wetland Designs**: Future urban wetland initiatives should account for rising temperatures, sea-level rise, and changing precipitation patterns.

Proactive planning and investment in wetland-friendly urban development will help cities address environmental challenges while improving residents' well-being.

Synergies with Green Infrastructure

Wetlands and green infrastructure work together to enhance urban resilience, improve water management, and support biodiversity. Green infrastructure refers to natural or engineered systems that mimic ecological processes to provide environmental, economic, and social benefits. By integrating wetlands with green infrastructure, cities can create sustainable solutions for flood control, water purification, climate adaptation, and urban livability. However, maximizing these synergies requires thoughtful planning, investment, and policy support to ensure wetlands and green infrastructure function together effectively.

The Role of Wetlands in Green Infrastructure

Wetlands contribute to green infrastructure by enhancing natural water regulation, improving air quality, and providing habitat for wildlife. Their integration into urban environments helps cities address challenges related to:

- **Stormwater Management**: Wetlands absorb and slow rainwater runoff, reducing the risk of urban flooding.
- **Water Filtration and Pollution Control**: Wetland plants and soils filter pollutants from stormwater and industrial discharge.
- **Heat Island Reduction**: Wetlands and associated green spaces cool urban areas by retaining moisture and providing shade.
- **Carbon Sequestration and Climate Mitigation**: Wetlands store carbon in vegetation and soils, reducing greenhouse gas emissions.
- **Biodiversity Conservation**: Urban wetlands support diverse plant and animal species, contributing to ecological balance.

By incorporating wetlands into green infrastructure strategies, cities can enhance their sustainability while maintaining natural ecosystem functions.

Key Synergies Between Wetlands and Green Infrastructure

Integrating wetlands with green infrastructure enhances urban resilience and supports climate adaptation efforts. These synergies can be achieved through various urban planning and design strategies.

Stormwater Management and Flood Prevention

One of the most critical synergies between wetlands and green infrastructure is stormwater regulation. As urban areas expand, impermeable surfaces like roads and buildings prevent water infiltration, leading to flooding and water pollution. Wetlands, combined with green infrastructure solutions, help mitigate these effects.

- **Bioswales and Constructed Wetlands**: Vegetated drainage channels and artificial wetlands capture and filter stormwater before it enters waterways.
- **Retention Ponds and Rain Gardens**: These features collect excess rainwater, reducing runoff and supporting groundwater recharge.
- **Permeable Pavements and Green Roofs**: These solutions work alongside wetlands to slow and absorb stormwater in urban settings.

By integrating wetlands with green stormwater infrastructure, cities can enhance flood resilience while maintaining clean water supplies.

Water Quality Improvement and Pollution Control

Wetlands naturally filter water, making them an essential component of urban green infrastructure aimed at improving water quality. When combined with engineered solutions, wetlands become even more effective in removing pollutants.

- **Constructed Treatment Wetlands**: These artificial wetlands are designed to process stormwater, wastewater, and industrial runoff by removing nutrients and heavy metals.
- **Wetland Buffer Zones**: Vegetated wetland edges prevent pollutants from reaching rivers and lakes, improving overall water quality.
- **Urban Green Corridors**: Integrating wetlands into park systems ensures continuous filtration and reduces contamination in urban waterways.

These synergies contribute to public health by ensuring cleaner water for cities while preserving aquatic ecosystems.

Enhancing Urban Cooling and Air Quality

The urban heat island effect raises temperatures in cities due to increased concrete surfaces and reduced vegetation. Wetlands, in combination with green infrastructure, help regulate temperatures and improve air quality.

- **Cooling Effects of Wetlands**: Water bodies and wetland vegetation reduce surrounding temperatures by evaporative cooling.
- **Integration with Urban Forests**: Combining wetlands with tree-lined streets and green roofs enhances shading and cooling benefits.
- **Air Filtration**: Wetland plants absorb pollutants such as nitrogen oxides and carbon dioxide, improving air quality in urban areas.

By integrating wetlands into urban green spaces, cities can mitigate heat stress, reduce energy consumption, and improve overall air quality.

Biodiversity and Habitat Connectivity

Urban expansion often fragments habitats, making it difficult for wildlife to move and thrive. Wetlands, when connected with green infrastructure, create biodiversity corridors that support species movement and ecological health.

- **Greenways and Riparian Buffers**: These natural corridors link wetlands with parks and forests, allowing wildlife to migrate and adapt to urban environments.
- **Pollinator and Bird Habitats**: Wetland plants attract birds, bees, and butterflies, enhancing biodiversity and supporting pollination services.
- **Multi-Use Wetland Parks**: Designing urban wetlands as public spaces ensures both ecological conservation and community engagement.

Ensuring connectivity between wetlands and other green infrastructure elements helps maintain species diversity while enhancing recreational opportunities for urban populations.

Challenges in Integrating Wetlands with Green Infrastructure

Despite their benefits, several challenges hinder the effective integration of wetlands into green infrastructure planning.

- **Competing Land Uses**: Urban expansion prioritizes development over wetland conservation, reducing available space for integration.
- **Water Management Conflicts**: Some cities over-rely on engineered water solutions, neglecting natural wetland systems.
- **Pollution and Habitat Degradation**: Poor wastewater management and industrial runoff can damage wetlands, limiting their ability to function effectively.
- **Lack of Awareness and Policy Support**: Many urban planning frameworks do not recognize wetlands as key components of green infrastructure.

Addressing these challenges requires stronger regulatory frameworks, interdisciplinary collaboration, and increased investment in NbS.

Strategies for Enhancing Wetland-Green Infrastructure Synergies

To maximize the benefits of wetlands in green infrastructure planning, cities can adopt the following strategies:

- **Incorporating Wetlands into Urban Master Plans**: Recognizing wetlands as critical infrastructure components ensures their protection and integration into city designs.
- **Expanding Funding for NbS**: Governments and private sectors should invest in wetland restoration and maintenance to support urban resilience.
- **Strengthening Policy and Legal Protections**: Establishing conservation policies that prioritize wetlands in green infrastructure projects safeguards their ecological functions.
- **Engaging Communities in Wetland Stewardship**: Public education and participatory planning encourage local support for wetland-friendly urban initiatives.

By aligning wetland conservation with urban sustainability goals, cities can create long-term environmental, social, and economic benefits.

Challenges and Opportunities

Integrating wetlands into urban landscapes presents both challenges and opportunities for sustainable city planning and environmental conservation. While wetlands provide essential ecosystem services such as flood control, water purification, and biodiversity support, they face significant pressures from urban expansion, pollution, and competing land uses. At the same time, growing recognition of NbS offers new opportunities to enhance wetland conservation while addressing urban challenges. By understanding these challenges and leveraging available opportunities, cities can better integrate wetlands into their planning frameworks and improve long-term environmental resilience.

Key Challenges in Wetland Conservation and Urban Integration

Despite their importance, wetlands face multiple threats that hinder their effective conservation and integration into urban settings. These challenges stem from land-use conflicts, pollution, policy gaps, and climate change impacts.

Competing Land-Use Priorities

One of the biggest challenges in urban wetland conservation is the pressure from urban expansion and infrastructure development. As cities grow, wetlands are often drained or converted into residential, commercial, or industrial areas.

- **Urbanization and Real Estate Development**: Wetlands are frequently seen as "undeveloped land," leading to their destruction for roads, housing, and business districts.
- **Infrastructure Projects**: Large-scale infrastructure, such as highways and airports, often leads to wetland fragmentation and loss.
- **Lack of Space for Restoration**: In heavily urbanized areas, restoring wetlands is difficult due to limited land availability and competing interests.

Addressing these conflicts requires stronger land-use planning that prioritizes wetland conservation alongside urban growth.

Pollution and Water Contamination

Urban wetlands are highly vulnerable to pollution from industrial discharge, stormwater runoff, and untreated wastewater.

Contaminants entering wetlands reduce their ability to filter water and support biodiversity.

- **Stormwater Runoff**: Urban surfaces, such as roads and buildings, produce runoff that carries oil, chemicals, and heavy metals into wetlands.
- **Agricultural and Industrial Pollution**: Pesticides, fertilizers, and industrial waste degrade wetland water quality, harming aquatic ecosystems.
- **Plastic and Microplastic Contamination**: Urban waste, including plastics, often accumulates in wetland areas, disrupting wildlife and ecological functions.

Implementing stricter pollution controls and expanding green infrastructure can help mitigate these threats.

Fragmentation and Loss of Connectivity

Urban development often isolates wetlands, making them smaller and disconnected from other ecosystems. This fragmentation reduces biodiversity, disrupts species migration, and weakens ecosystem resilience.

- **Loss of Wildlife Corridors**: Species that depend on wetlands for breeding and migration struggle to survive when habitats are fragmented.
- **Disrupted Hydrological Cycles**: When wetlands lose connection to natural waterways, they can no longer regulate floods or support groundwater recharge effectively.
- **Decline in Ecosystem Function**: Small, isolated wetlands are less effective in filtering pollutants, storing carbon, and maintaining biodiversity.

Urban planners must focus on ecological connectivity, linking wetlands with other green spaces through green corridors and habitat restoration.

Weak Policies and Lack of Enforcement

Many cities lack comprehensive wetland protection laws, and existing policies are often poorly enforced.

- **Inconsistent Land-Use Regulations**: Some regions have conflicting laws that allow wetland destruction for economic development.
- **Limited Funding for Conservation**: Governments and municipalities often underfund wetland conservation programs, prioritizing short-term infrastructure projects instead.
- **Weak Enforcement Mechanisms**: Even where wetland protection laws exist, they are frequently ignored due to weak monitoring and enforcement.

Stronger governance, policy integration, and financial incentives are needed to ensure wetland protection is a priority in urban planning.

Climate Change and Extreme Weather Events

Climate change is altering rainfall patterns, increasing temperatures, and intensifying extreme weather events, all of which threaten urban wetlands.

- **Rising Sea Levels**: Coastal wetlands, such as mangroves and salt marshes, are at risk of being submerged.
- **Prolonged Droughts**: Reduced rainfall lowers water levels in urban wetlands, affecting their ability to store water and support biodiversity.
- **Intensified Flooding**: Extreme rainfall events can overwhelm urban wetlands, causing erosion and habitat degradation.

Adapting to climate change requires enhanced wetland restoration, improved flood management, and climate-resilient urban planning.

Opportunities for Enhancing Wetland Conservation in Urban Areas

Despite these challenges, there are significant opportunities to integrate wetlands into urban landscapes through policy innovation, green infrastructure, and community engagement.

Advancing NbS in Urban Planning

Wetlands can be integrated into sustainable urban development strategies to enhance climate resilience and improve quality of life.

- **Constructed Wetlands for Stormwater Treatment**: Cities can develop engineered wetlands that filter pollutants and manage urban runoff.
- **Green Infrastructure Expansion**: Combining wetlands with bioswales, green roofs, and rain gardens improves stormwater management.
- **Urban Wetland Parks**: Designing wetlands as multi-functional spaces provides both ecological and recreational benefits.

By promoting NbS, cities can enhance wetland protection while addressing urban challenges.

Strengthening Wetland Protection Policies

Governments and municipalities can implement stronger policies and incentives to ensure wetlands are conserved.

- **Integrating Wetlands into Urban Zoning Laws**: Designating wetlands as protected areas prevents destructive land-use changes.
- **Creating Financial Incentives for Conservation**: Tax benefits and subsidies encourage landowners and businesses to preserve wetlands.

- **Improving Enforcement and Monitoring**: Investing in remote sensing and community-based monitoring ensures better compliance with conservation laws.

Stronger legal frameworks can help secure wetlands as essential urban infrastructure.

Engaging Communities in Wetland Stewardship

Public participation in wetland conservation fosters long-term sustainability and stronger policy support.

- **Community-Led Restoration Projects**: Local groups can help restore degraded wetlands through native planting and pollution cleanup.
- **Environmental Education and Awareness**: Schools, businesses, and local organizations can promote wetland conservation through awareness campaigns.
- **Citizen Science and Monitoring Programs**: Engaging the public in wetland health monitoring improves conservation outcomes and strengthens community involvement.

Encouraging local stewardship ensures that urban wetlands remain valued and protected by the people who benefit from them.

Leveraging Public-Private Partnerships

Collaboration between government agencies, businesses, and non-profit organizations can expand resources for wetland conservation.

- **Corporate Sustainability Programs**: Businesses can support wetland conservation through corporate responsibility initiatives.
- **Infrastructure Projects with Environmental Offsets**: Developers can be required to fund wetland restoration to compensate for land use impacts.

- **Innovative Financing Models**: Carbon credits and wetland conservation bonds can provide funding for restoration projects.

Public-private partnerships enhance funding opportunities while aligning conservation with economic development goals.

Chapter 8: Economics of Wetlands as Nature-Based Solutions

Wetlands provide essential ecosystem services that support economies, enhance climate resilience, and contribute to sustainable development. As NbS, wetlands offer cost-effective alternatives to engineered infrastructure for flood protection, water purification, and carbon sequestration. However, despite their economic value, wetlands are often undervalued in financial and policy decisions, leading to their degradation and loss.

Understanding the economic benefits of wetlands is crucial for integrating them into decision-making, securing funding for conservation, and incentivizing sustainable management. This chapter explores the economic role of wetlands, including their contributions to local and global economies, cost-benefit comparisons with artificial infrastructure, and financial mechanisms that support wetland conservation and restoration. It also examines challenges in valuing wetlands and the opportunities for aligning wetland protection with sustainable economic growth.

Economic Valuation of Wetland Services

Wetlands provide essential ecosystem services that support economies, enhance resilience, and contribute to human well-being. However, these benefits are often undervalued or overlooked in economic decision-making, leading to wetland degradation and loss. Economic valuation helps quantify the benefits wetlands provide, demonstrating their financial importance and making a case for their conservation. By assigning a monetary value to wetland services, policymakers, businesses, and communities can make informed decisions that balance development with sustainability.

Why Economic Valuation of Wetlands Matters

Wetland ecosystems contribute to economies in several ways, yet they are often treated as unproductive land. Economic valuation is necessary to:

- **Highlight the True Value of Wetlands**: Many benefits, such as flood protection and carbon sequestration, are not directly traded in markets and remain invisible in economic calculations.
- **Justify Conservation Investments**: Assigning a value to wetland services helps secure funding for conservation and restoration efforts.
- **Inform Policy and Land-Use Decisions**: Understanding the economic impact of wetland loss can influence policies that balance development with environmental protection.
- **Encourage Sustainable Business Practices**: Industries that depend on wetland services, such as fisheries, tourism, and water utilities, can use valuation data to promote sustainable resource use.

Economic valuation ensures that wetland benefits are recognized and integrated into decision-making processes.

Methods for Valuing Wetland Ecosystem Services

Several economic approaches are used to assess the value of wetlands, depending on the type of service and available data.

Direct Market Valuation

Some wetland services have direct market values because they generate income or involve measurable transactions.

- **Fisheries and Aquaculture**: Wetlands support commercial and subsistence fisheries, providing direct financial benefits to local and global markets.

- **Agricultural Productivity**: Floodplain wetlands contribute to soil fertility, reducing the need for fertilizers and enhancing crop yields.
- **Timber and Non-Timber Forest Products**: Mangroves and swamp forests provide wood, fuel, and medicinal plants that are sold in markets.

This method applies to goods and services that have established market prices, making it relatively easy to quantify.

Replacement Cost Method

This approach estimates the cost of replacing wetland services with artificial infrastructure. It highlights how wetlands offer cost-effective solutions compared to built alternatives.

- **Flood Protection**: Wetlands act as natural flood buffers, reducing the need for levees and drainage systems. The cost of replacing wetland flood control with artificial infrastructure can be substantial.
- **Water Filtration**: Wetlands remove pollutants from water, reducing the need for expensive water treatment plants. The replacement cost is the price of engineered filtration systems.
- **Coastal Protection**: Mangroves and salt marshes reduce storm surge impacts, lowering the need for costly seawalls and barriers.

The replacement cost method emphasizes how wetlands provide free or low-cost services that would otherwise require large financial investments.

Travel Cost Method (Tourism and Recreation Value)

Wetlands attract visitors for ecotourism, birdwatching, fishing, and outdoor recreation. The travel cost method estimates their value based on how much people are willing to spend to visit them.

- **Entrance Fees and Tourism Revenue**: Many wetlands are part of national parks or protected areas that generate income from tourism.
- **Accommodation and Local Business Revenue**: Hotels, restaurants, and tour operators benefit from wetland-related tourism.
- **Cultural and Educational Value**: Wetlands provide educational and recreational experiences that contribute to well-being and local economies.

This approach highlights the economic importance of wetland-based tourism and its potential for sustainable development.

Contingent Valuation (Willingness to Pay for Conservation)

This method assesses how much people are willing to pay to protect or restore wetlands, even if they do not directly use them.

- **Public Support for Conservation Programs**: Surveys can estimate how much individuals or businesses are willing to contribute to wetland restoration.
- **Non-Use Values**: Many people value wetlands for their existence and cultural significance, even if they do not visit them.

This approach helps quantify intangible benefits, such as biodiversity preservation and cultural heritage.

Carbon Sequestration Valuation

Wetlands, particularly peatlands, mangroves, and salt marshes, store large amounts of carbon, helping mitigate climate change. The value of carbon sequestration can be estimated using:

- **Carbon Market Prices**: Wetlands can generate carbon credits that industries buy to offset emissions.

- **Social Cost of Carbon**: The economic damage caused by carbon emissions provides a measure of how much wetlands contribute to climate change mitigation.

Recognizing wetlands as carbon sinks strengthens their economic justification for conservation.

Challenges in Wetland Valuation

Despite its benefits, economic valuation of wetlands faces several challenges:

- **Lack of Market Prices for Many Services**: Many wetland benefits, such as flood regulation and biodiversity, do not have direct financial transactions, making valuation complex.
- **Data Limitations and Uncertainty**: Measuring wetland services accurately requires extensive ecological and economic data, which is often unavailable.
- **Short-Term Economic Pressures**: Many development projects prioritize immediate financial gains over long-term wetland benefits.
- **Difficulties in Monetizing Cultural and Non-Use Values**: Cultural and intrinsic values of wetlands are hard to quantify in purely financial terms.

Overcoming these challenges requires multidisciplinary collaboration, better data collection, and greater awareness of wetland benefits in economic planning.

Future Opportunities for Wetland Economic Valuation

Advancing wetland valuation can support stronger conservation policies and sustainable financing mechanisms.

- **Integrating Wetlands into National Accounting**: Governments can include wetland services in Gross Domestic Product (GDP) and natural capital accounting.

- **Developing PES Programs**: Landowners and businesses can be compensated for wetland conservation.
- **Expanding Carbon Markets**: Wetlands can play a bigger role in carbon offset programs, attracting investment for restoration projects.
- **Using Technology for Better Valuation**: Remote sensing, artificial intelligence, and economic modeling can improve wetland valuation accuracy.

Recognizing the economic value of wetlands can drive better policies, increase funding for conservation, and promote sustainable development.

Comparing Costs with Engineered Solutions

Wetlands provide cost-effective NbS for managing water, mitigating climate impacts, and protecting communities from environmental hazards. However, many cities and industries rely on engineered infrastructure, such as levees, dams, and water treatment plants, to address similar challenges. While built infrastructure can deliver immediate and predictable outcomes, it often involves high construction, maintenance, and operational costs. By contrast, wetlands offer long-term, self-sustaining benefits at a fraction of the cost. Comparing the financial investment required for engineered solutions with the services wetlands provide demonstrates the economic value of conserving and restoring these ecosystems.

The Cost of Engineered Infrastructure vs. Wetlands

Engineered solutions require substantial upfront investment, as well as continuous maintenance and upgrades. Wetlands, in contrast, require lower capital investment and provide ecosystem services that naturally regenerate over time.

Comparing Costs of Flood Protection

Engineered Solutions: Levees, Dams, and Drainage Systems

- **High Initial Construction Costs**: Levees and drainage systems require billions in investments, particularly in flood-prone cities.
- **Ongoing Maintenance**: Infrastructure deteriorates over time, requiring continuous repair, reinforcement, and upgrades.
- **Limited Climate Adaptation**: As climate change intensifies storms and sea-level rise, traditional flood defenses must be expanded or rebuilt at great expense.

For example, New Orleans' levee system, designed after Hurricane Katrina, cost over $14 billion and requires regular maintenance costing millions annually.

Nature-Based Solution: Wetlands for Flood Control

- **Cost Savings on Infrastructure**: Coastal wetlands absorb up to 30% of storm surge energy, reducing the need for high-cost barriers.
- **Self-Sustaining and Adaptive**: Unlike levees, wetlands evolve over time, naturally adjusting to rising water levels.
- **Long-Term Resilience**: Restored floodplain wetlands lower the risk of downstream flooding, reducing future financial losses.

Studies estimate that one hectare of coastal wetland can provide up to $33,000 per year in flood protection benefits—a fraction of the cost of engineered flood defenses.

Comparing Costs of Water Filtration and Treatment

Engineered Solutions: Water Treatment Plants

- **Energy-Intensive Processes**: Urban water treatment facilities require electricity, chemicals, and regular maintenance.

- **Expensive Construction and Operation**: Large-scale water treatment plants cost millions to billions to build and maintain.
- **Limited Pollution Reduction**: Chemical treatment processes do not always remove all contaminants, especially emerging pollutants like pharmaceuticals and microplastics.

Nature-Based Solution: Wetlands as Natural Water Filters

- **Natural Filtration**: Wetlands trap sediments, remove excess nutrients, and break down pollutants without additional energy inputs.
- **Lower Long-Term Costs**: Instead of requiring upgrades, wetlands continue providing filtration services indefinitely with minimal intervention.
- **Multiple Benefits Beyond Water Quality**: In addition to filtering water, wetlands enhance biodiversity and store carbon.

For example, New York City avoided $8 billion in new water treatment infrastructure costs by investing in wetland conservation in the Catskill Watershed, ensuring high-quality drinking water at lower costs.

Comparing Costs of Storm Surge Protection

Engineered Solutions: Seawalls and Coastal Barriers

- **High Construction Costs**: Seawalls and breakwaters can cost millions per kilometer, depending on location and material.
- **Expensive Repairs After Storms**: Extreme weather events can cause structural damage, requiring costly reinforcement.
- **Ecological Damage**: Concrete structures disrupt coastal ecosystems, accelerating erosion and biodiversity loss.

For instance, the seawall protecting Galveston, Texas, built after the 1900 hurricane, requires continuous upgrades costing millions annually.

Nature-Based Solution: Mangroves and Coastal Wetlands

- **Significant Wave Energy Reduction**: Mangroves and salt marshes can reduce storm surge height by up to 2 meters over short distances.
- **Lower Costs, Higher Resilience**: Restored coastal wetlands are cheaper and more effective in the long term than artificial barriers.
- **Biodiversity and Fisheries Benefits**: Coastal wetlands also support fisheries, tourism, and local economies while providing protection.

A study found that mangroves in the Philippines prevent $1.6 billion in annual storm damage, a service that would require extensive infrastructure investment to replicate.

Challenges of Relying Solely on Wetlands as NbS

Despite their cost advantages, wetlands alone cannot replace all engineered solutions. Some challenges include:

- **Time Required for Restoration**: Wetland recovery takes years, while engineered solutions provide immediate benefits.
- **Space Requirements**: Large wetland areas are needed to match the effectiveness of engineered infrastructure, which is challenging in dense urban areas.
- **Uncertain Policy Support**: Many governments prioritize short-term infrastructure projects over long-term ecosystem investments.

To maximize benefits, hybrid approaches that combine natural and engineered solutions—such as green infrastructure integrating wetlands with stormwater retention basins—offer the best outcomes.

Opportunities for Increasing NbS Adoption

Governments, businesses, and communities can invest in wetlands as low-cost, high-impact alternatives to traditional infrastructure. Key strategies include:

- **Incentivizing Wetland Restoration**: PES and carbon markets can fund wetland conservation.
- **Incorporating Wetlands into Urban Planning**: Zoning laws should protect and integrate wetlands into flood-prone areas.
- **Blended Approaches**: Combining wetlands with stormwater retention ponds, levees, and bioswales enhances resilience while reducing costs.
- **Corporate and Public Investment**: Businesses can offset carbon footprints through wetland conservation projects, while governments can redirect infrastructure budgets toward NbS.

Funding Mechanisms

Securing adequate funding is essential for the conservation, restoration, and sustainable management of wetlands as NbS. Despite their significant environmental and economic benefits, wetlands are often undervalued in traditional financial planning, leading to underfunding and degradation. Effective funding mechanisms can help ensure that wetlands receive the necessary resources to support climate resilience, biodiversity conservation, and water management.

This section explores different funding approaches, including public investments, private sector engagement, market-based solutions, and international funding initiatives, to support wetlands as NbS. By diversifying funding sources, stakeholders can create sustainable financial strategies that promote long-term wetland conservation and restoration.

Public Sector Funding

Governments play a crucial role in financing wetland conservation through national budgets, grants, and incentive programs. Public funding ensures that wetland conservation remains a priority in national and regional environmental policies.

Government Grants and Subsidies

- **Conservation and Restoration Grants**: Many governments allocate funds for wetland restoration projects, biodiversity protection, and water management programs.
- **Subsidies for Sustainable Land Use**: Farmers and landowners receive subsidies for preserving wetlands on their property instead of converting them for agriculture or development.
- **Disaster Risk Reduction Funding**: Governments invest in wetlands as flood protection and climate adaptation strategies, reducing costs associated with storm damage and infrastructure repair.

Public investments in wetlands as natural infrastructure can lower long-term spending on engineered flood control, water treatment, and disaster recovery.

Tax Incentives for Wetland Conservation

- **Property Tax Reductions**: Landowners who conserve or restore wetlands on their property can qualify for lower tax rates.
- **Tax Credits for Restoration Projects**: Businesses and individuals investing in wetland conservation efforts can receive tax benefits.

- **Green Infrastructure Incentives**: Cities and municipalities provide financial rewards for developments that integrate wetland preservation into urban planning.

Tax incentives encourage private participation in wetland conservation while reducing the financial burden on governments.

Private Sector and Corporate Funding

Businesses and industries benefit from wetland ecosystem services, such as water purification, flood control, and carbon sequestration. Encouraging corporate investment in wetland conservation can generate funding while supporting corporate social responsibility (CSR) goals.

CSR Investments

- **Sustainability Initiatives**: Companies invest in wetland restoration as part of their environmental and sustainability commitments.
- **Offsets for Industrial Impacts**: Businesses that affect wetlands (e.g., developers, extractive industries) fund wetland restoration projects to compensate for environmental damage.
- **Branding and Eco-Certifications**: Companies that support wetland conservation can enhance their reputation and market appeal through sustainability certifications and green branding.

Public-Private Partnerships (PPPs)

- **Joint Conservation Programs**: Governments and businesses co-finance wetland protection to achieve mutual environmental and economic benefits.
- **Infrastructure Offsetting Agreements**: Companies developing industrial projects, highways, or urban expansion invest in wetland restoration as a compensation measure.

- **Water Utility Partnerships**: Water companies fund wetland conservation to ensure sustainable water supplies and reduce treatment costs.

PPP models leverage private sector innovation and funding to expand conservation efforts beyond traditional government budgets.

Market-Based Mechanisms

Market-driven solutions provide financial incentives for wetland conservation by linking ecosystem services with economic markets and financial products.

PES

PES programs compensate individuals or communities for preserving wetlands and maintaining ecosystem services.

- **Water Quality Trading Programs**: Polluters pay for wetland conservation that helps offset water pollution impacts.
- **Biodiversity Credits**: Developers invest in wetland restoration projects to offset habitat destruction elsewhere.
- **Carbon Markets and Wetland-Based Carbon Credits**: Wetlands store significant amounts of carbon, and peatland and mangrove restoration projects can generate carbon credits for trading in emissions markets.

PES models turn wetland conservation into an economic opportunity, attracting private sector participation.

Wetland Conservation Bonds

- **Green Bonds**: These bonds finance wetland restoration, flood management, and water security projects, with investors receiving returns based on project success.

- **Blue Bonds**: Specifically designed for coastal and marine wetland conservation, blue bonds fund mangrove and salt marsh restoration.
- **Impact Investment Funds**: Investors allocate capital to wetland conservation projects that generate measurable environmental and social benefits.

Bonds offer long-term, sustainable financing for wetland protection while appealing to environmentally conscious investors.

International and Multilateral Funding

Global institutions and environmental funds provide grants and financial assistance for wetland conservation in developing regions and climate-vulnerable areas.

Multilateral Environmental Funds

- **Global Environment Facility (GEF)**: Provides grants for wetland biodiversity conservation and climate resilience projects.
- **Green Climate Fund (GCF)**: Supports wetland-based climate adaptation and mitigation efforts in developing countries.
- **Ramsar Convention Small Grants Program**: Funds wetland restoration initiatives worldwide under the Ramsar Convention on Wetlands.

These international funds support large-scale wetland projects, particularly in areas facing severe climate and water security risks.

NGOs and Philanthropic Support

- **Conservation NGOs**: Organizations such as WWF, The Nature Conservancy, and Wetlands International provide funding and expertise for wetland conservation.

- **Private Foundations and Donors**: Environmental philanthropists and foundations fund community-led wetland restoration projects.

NGOs and philanthropic contributions play a critical role in financing wetland conservation where government support is limited.

Challenges in Securing Sustainable Wetland Funding

Despite the variety of funding mechanisms, several challenges hinder consistent financial support for wetlands:

- **Lack of Policy Integration**: Many national budgets prioritize infrastructure development over wetland conservation, limiting government funding.
- **Short-Term Financial Planning**: Conservation requires long-term investment, but many funding sources operate on short project cycles.
- **Limited Private Sector Participation**: Businesses often lack awareness of how wetland conservation aligns with corporate sustainability goals.
- **Weak Financial Incentives**: Many regions lack strong economic incentives for individuals and companies to invest in wetlands.

Strengthening Funding Strategies for Wetland Conservation

To overcome funding challenges, stakeholders can adopt integrated financial strategies:

- **Mainstream Wetlands into National Budgets**: Governments should include wetland conservation in infrastructure, water management, and climate adaptation budgets.

- **Expand Market-Based Incentives**: Strengthening carbon markets, PES programs, and conservation bonds can attract private investment.
- **Develop Long-Term Public-Private Partnerships**: Collaboration between governments, businesses, and NGOs ensures continuous funding.
- **Increase Public Awareness on Wetland Economics**: Educating stakeholders about the financial benefits of wetlands can drive investment and policy support.

Chapter 9: Wetlands in Policy and Governance

Wetlands play a crucial role in water security, climate resilience, and biodiversity conservation, yet they remain vulnerable to degradation due to weak policies, competing land uses, and insufficient enforcement. Effective policy and governance frameworks are essential for protecting wetlands, ensuring sustainable management, and integrating them into broader environmental and economic planning.

Governments, international organizations, and local stakeholders all play a role in wetland governance, with policies ranging from global agreements like the Ramsar Convention to national and regional wetland protection laws. However, challenges such as fragmented policies, lack of enforcement, and limited funding often hinder conservation efforts.

This chapter explores the role of policy and governance in wetland conservation, examining international frameworks, national regulations, community participation, and challenges in implementation. It highlights strategies for strengthening wetland policies, improving enforcement, and integrating wetlands into sustainable development planning.

International Conventions

Wetlands are globally significant ecosystems that provide water security, climate resilience, and biodiversity conservation. However, their transboundary nature and vulnerability to degradation require international cooperation and governance frameworks to ensure their protection. Several international conventions and agreements have been established to support wetland conservation, promote sustainable management, and encourage cross-border collaboration. These conventions play a crucial role in setting global standards, providing funding mechanisms, and guiding national policies to safeguard wetlands.

This section explores key international conventions relevant to wetland protection, their objectives, implementation mechanisms, and challenges in enforcement.

Ramsar Convention on Wetlands

Overview

The Ramsar Convention on Wetlands (adopted in 1971 in Ramsar, Iran) is the first and only global treaty dedicated specifically to wetlands. It aims to ensure the conservation and sustainable use of wetlands worldwide by recognizing their ecological, economic, and cultural importance.

Key Objectives

- **Identifying and Protecting Wetlands of International Importance**: Member countries designate at least one Wetland of International Importance (Ramsar Site) and commit to its protection.
- **Sustainable Use of All Wetlands**: Encourages wise use of wetlands through national planning and legislation.
- **International Cooperation**: Promotes transboundary wetland management and information sharing between countries.

Implementation Mechanisms

- **Ramsar Site Designation**: Over 2,400 wetlands across 170+ countries have been designated as Ramsar Sites, ensuring global recognition and protection.

- **National Wetland Policies**: Countries integrate wetland conservation into land-use planning and water management policies.
- **Financial Support**: The Ramsar Small Grants Fund (SGF) provides funding for wetland conservation projects, particularly in developing countries.

Challenges in Enforcement

- **Lack of Legal Binding Power**: Unlike other environmental agreements, Ramsar does not impose legal sanctions for non-compliance.
- **Limited Funding for Conservation**: Many Ramsar-listed wetlands lack sufficient financial and technical resources for long-term protection.
- **Competing Land-Use Pressures**: Economic activities such as agriculture, infrastructure development, and urban expansion continue to threaten Ramsar Sites.

Despite these challenges, the Ramsar Convention remains a cornerstone of global wetland governance, influencing national policies and conservation strategies.

Convention on Biological Diversity (CBD)

Overview

The CBD, adopted at the 1992 Earth Summit in Rio de Janeiro, aims to protect global biodiversity, including wetland ecosystems that support species diversity and ecological resilience.

Key Objectives

- **Conserving Biodiversity**: Wetlands are recognized as critical habitats for countless species, including migratory birds, amphibians, and fish.

- **Sustainable Use of Ecosystems**: Encourages integrating wetlands into national biodiversity strategies and action plans (NBSAPs).
- **Equitable Sharing of Benefits**: Supports community-based conservation and fair use of wetland resources.

Implementation Mechanisms

- **Global Biodiversity Framework**: Countries develop national action plans to conserve wetlands and integrate them into land-use policies.
- **Protected Area Networks**: Encourages expansion of wetland reserves and conservation areas.
- **Funding and Technical Assistance**: The GEF provides financial support for wetland conservation projects.

Challenges in Enforcement

- **Weak Legal Enforcement**: Compliance relies on voluntary national commitments rather than legally binding measures.
- **Fragmented Implementation**: Wetlands often fall under multiple government agencies, leading to overlapping or ineffective policies.
- **Biodiversity Loss Due to Climate Change**: Wetland species face growing threats from rising temperatures, habitat destruction, and pollution.

United Nations Framework Convention on Climate Change (UNFCCC)

Overview

The UNFCCC, adopted in 1992, addresses climate change impacts, including those affecting wetlands. Wetlands play a key role in carbon sequestration, flood regulation, and climate adaptation, making them essential for global climate resilience strategies.

Key Objectives

- **Recognizing Wetlands as Carbon Sinks**: Peatlands, mangroves, and marshes store large amounts of carbon, contributing to global emissions reduction goals.
- **Enhancing Climate Adaptation**: Wetlands help reduce disaster risks from flooding and sea-level rise.
- **Funding for NbS**: Supports wetland conservation through mechanisms like the GCF.

Implementation Mechanisms

- **Nationally Determined Contributions (NDCs)**: Many countries include wetland restoration in their climate commitments.
- **REDD+ Program**: Provides incentives for protecting carbon-storing ecosystems like peatlands and mangroves.
- **Climate Adaptation Funds**: The GCF and Adaptation Fund support projects that use wetlands to mitigate climate risks.

Challenges in Enforcement

- **Limited Integration of Wetlands into Climate Policy**: Many climate action plans overlook wetlands in favor of technological solutions.
- **Funding Gaps**: Many developing countries struggle to access climate finance for wetland restoration.
- **Pressure from Development Projects**: Coastal wetlands face threats from urbanization, industrialization, and land reclamation despite their role in climate resilience.

Convention on Migratory Species (CMS)

Overview

The CMS, also known as the Bonn Convention (1979), protects species that depend on wetlands for breeding, feeding, and migration.

Key Objectives

- **Protecting Wetland-Dependent Species**: Many migratory birds, fish, and amphibians rely on healthy wetland ecosystems.
- **Establishing Transboundary Conservation Areas**: Encourages cross-border cooperation to protect migration routes.
- **Preventing Habitat Loss**: Works alongside Ramsar and CBD to prevent wetland degradation.

Implementation Mechanisms

- **Species-Specific Agreements**: Countries sign regional agreements to protect migratory species.
- **Monitoring and Data Sharing**: Supports research and conservation programs for migratory species.
- **Funding for Habitat Protection**: Encourages governments and NGOs to invest in wetland conservation initiatives.

Challenges in Enforcement

- **Weak Coordination Between Countries**: Migratory species cross multiple jurisdictions, requiring stronger international agreements.
- **Wetland Loss Along Migration Routes**: Many key wetlands are lost due to urbanization and climate change, disrupting migration patterns.
- **Limited Public Awareness**: Conservation efforts often lack visibility and funding compared to other environmental issues.

National and Local Policies

Wetlands provide essential ecosystem services, including flood control, water filtration, carbon sequestration, and biodiversity conservation. However, despite their ecological and economic value, wetlands remain vulnerable to urban expansion, pollution, and climate change. Effective national and local policies are critical for ensuring the protection, sustainable management, and restoration of wetlands. These policies set the legal and institutional frameworks for wetland conservation, guiding land-use planning, environmental regulations, and water resource management.

This section explores key national and local policy mechanisms, their role in wetland governance, and the challenges in enforcement and implementation.

National Wetland Protection Laws and Regulations

Many countries have national policies and legal frameworks dedicated to wetland conservation, often aligned with international agreements like the Ramsar Convention and the CBD.

Key National Policies

- **Wetland Conservation Acts**: Countries such as India, the United States, and Australia have enacted laws specifically designed to protect and regulate wetlands.
- **Environmental Protection Laws**: Wetlands are often covered under broader environmental legislation, such as the Clean Water Act in the United States or the Environmental Protection and Biodiversity Conservation Act in Australia.
- **Water Resource Management Policies**: Some nations integrate wetlands into water governance frameworks, recognizing their role in watershed health and flood mitigation.

These policies typically define wetland boundaries, permissible activities, and conservation measures, ensuring wetlands remain intact and functional.

Implementation Mechanisms

- **Legal Designation of Protected Wetlands**: National governments classify wetlands as protected areas, nature reserves, or Ramsar Sites to safeguard them from development.
- **Permitting and Regulatory Oversight**: Landowners and industries must obtain permits before altering or draining wetlands, ensuring compliance with environmental regulations.
- **Environmental Impact Assessments (EIA)**: Many policies require EIA approval before infrastructure or industrial projects near wetlands can proceed.

These mechanisms ensure wetlands are managed responsibly while allowing for sustainable development and land-use planning.

Local Government and Municipal Policies

Local governments play a crucial role in implementing wetland conservation policies at the community level, ensuring that national frameworks translate into effective action.

Key Local Policies

- **Land-Use Zoning Regulations**: Many cities and municipalities establish buffer zones around wetlands, limiting construction and industrial activity.
- **Stormwater and Drainage Management Plans**: Local policies incorporate wetlands into urban flood control systems, reducing reliance on costly artificial drainage.
- **Green Infrastructure Integration**: Cities integrate wetlands into urban parks, stormwater retention areas, and green spaces to enhance biodiversity and water management.
- **Community-Led Wetland Restoration Programs**: Some municipalities work with local communities and NGOs to

restore degraded wetlands through tree planting, invasive species removal, and ecological monitoring.

Implementation Mechanisms

- **Local Wetland Management Authorities**: Municipalities establish dedicated wetland conservation agencies or task forces to oversee wetland protection and restoration.
- **Sustainable Development Guidelines**: Local building codes incorporate NbS that prioritize wetland conservation, reducing the environmental impact of urban expansion.
- **Public Engagement and Education**: Local policies often emphasize awareness campaigns, citizen science programs, and school initiatives to promote wetland stewardship.

By integrating wetland protection into local planning frameworks, municipalities ensure that urban development aligns with ecological sustainability.

Challenges in National and Local Wetland Policies

Despite their importance, wetland policies at both national and local levels face significant implementation challenges.

Conflicting Land-Use Priorities

- **Urban Expansion and Infrastructure Projects**: Many wetlands are drained or filled to make way for housing, commercial developments, and roads, despite legal protections.
- **Agricultural and Industrial Pressures**: In rural areas, wetlands are often converted into farmland or used for industrial waste disposal, leading to habitat destruction and pollution.

Balancing economic development with wetland conservation remains a major policy challenge.

Weak Enforcement and Monitoring

- **Insufficient Government Oversight**: Many wetland laws exist on paper but lack proper enforcement due to weak institutional capacity.
- **Limited Funding for Conservation Programs**: Many local governments lack financial resources to implement wetland protection measures effectively.
- **Illegal Encroachments and Pollution**: Wetlands continue to suffer from unregulated construction, waste dumping, and water contamination, undermining conservation efforts.

Stronger policy enforcement and funding mechanisms are needed to ensure compliance and long-term wetland protection.

Fragmented Governance and Policy Coordination

- **Multiple Agencies Overseeing Wetlands**: In many countries, wetlands fall under multiple government agencies, leading to conflicting policies and inefficiencies.
- **Lack of Integrated Water and Land-Use Planning**: Many wetland policies operate in isolation rather than being integrated into broader climate adaptation, disaster risk reduction, and water resource management frameworks.
- **Limited Stakeholder Collaboration**: Effective wetland governance requires cooperation between governments, businesses, and local communities, which is often lacking.

A **more integrated and cross-sectoral approach** to wetland policy-making is needed to address these governance gaps.

Opportunities for Strengthening National and Local Wetland Policies

To improve wetland conservation and management, governments at all levels can adopt stronger policies, better enforcement mechanisms, and community-driven approaches.

Strengthening Policy Integration

- **Embedding Wetland Conservation into National Development Plans**: Governments should include wetlands in national economic, infrastructure, and water management strategies.
- **Developing Cross-Sectoral Policies**: Wetland policies should align with climate adaptation, disaster resilience, and agriculture policies to create a holistic governance framework.

Expanding Financial and Market-Based Incentives

- **PES**: Governments can offer financial rewards to landowners who preserve or restore wetlands.
- **Carbon Markets and Wetland Restoration Funding**: Protecting carbon-rich wetlands like **peatlands and mangroves** can attract climate finance.
- **Corporate Partnerships for Conservation**: Public-private partnerships can mobilize resources for local wetland management initiatives.

Enhancing Public Participation and Community Involvement

- **Community-Based Wetland Management**: Engaging local communities in wetland restoration and monitoring ensures long-term stewardship.
- **Public Education and Awareness Campaigns**: Schools, businesses, and community organizations can promote wetland conservation through outreach programs.
- **Citizen Science and Wetland Monitoring**: Encouraging community participation in data collection and policy advocacy strengthens conservation efforts.

Community and Stakeholder Engagement

Community and stakeholder engagement is essential for the effective conservation, restoration, and sustainable management of wetlands. While governments and international organizations play key roles in setting policies and regulations, local communities, businesses, indigenous groups, and civil society organizations are directly impacted by wetland management decisions. Their participation ensures that conservation efforts are not only effective but also socially and economically inclusive.

Engaging stakeholders in wetland management promotes long-term stewardship, improves policy outcomes, and strengthens public support for wetland protection. This section explores the importance of community involvement, stakeholder collaboration, challenges in engagement, and strategies for inclusive wetland governance.

The Importance of Community Engagement in Wetland Conservation

Wetlands provide essential services to local communities, including clean water, food, flood protection, and climate resilience. However, many communities remain underrepresented in decision-making processes, leading to conservation strategies that may overlook local needs and priorities.

Benefits of Community Engagement

- **Stronger Wetland Stewardship**: When local communities are actively involved, they are more likely to protect and restore wetlands rather than degrade them.
- **Integration of Local Knowledge**: Indigenous and rural communities often have deep ecological knowledge about wetland ecosystems, water cycles, and species interactions.
- **Economic and Social Benefits**: Community-led conservation programs can generate sustainable livelihoods through ecotourism, fisheries, and wetland-compatible agriculture.

- **Public Awareness and Support**: Educating local communities about wetland benefits encourages greater participation in conservation efforts.

Engaging local populations in wetland management enhances ecological, social, and economic sustainability, ensuring that conservation efforts align with community priorities.

Key Stakeholders in Wetland Management

Successful wetland governance requires input from a diverse range of stakeholders. Each group has different interests, responsibilities, and contributions to wetland conservation.

Local Communities and Indigenous Groups

- Primary users of wetland resources (e.g., fishers, farmers, traditional healers, and small-scale businesses).
- Hold valuable traditional ecological knowledge that contributes to sustainable wetland use.
- Often depend on wetlands for livelihoods, water supply, and cultural practices.

Government Agencies

- Develop and implement wetland protection laws, zoning regulations, and conservation programs.
- Oversee EIAs for projects affecting wetlands.
- Provide funding, technical expertise, and enforcement of conservation regulations.

Private Sector and Businesses

- Agriculture, real estate, tourism, and industry sectors often have economic interests in wetland areas.
- Responsible for mitigating their environmental impact through sustainable business practices.

- Can invest in CSR initiatives that support wetland conservation.

NGOs and Conservation Groups

- Conduct scientific research, wetland restoration, and public awareness campaigns.
- Advocate for stronger wetland policies and community rights.
- Support local capacity-building and education initiatives.

International Organizations and Donor Agencies

- Provide technical and financial support for wetland conservation programs.
- Develop global policies through international conventions (e.g., Ramsar Convention, UNFCCC).
- Promote transboundary wetland cooperation between countries.

Each stakeholder group brings unique expertise, resources, and perspectives, making collaborative governance essential for effective wetland conservation.

Challenges in Community and Stakeholder Engagement

Despite its importance, engaging communities and stakeholders in wetland management faces several obstacles.

Lack of Awareness and Education

- Many local communities lack knowledge about wetland benefits, leading to low participation in conservation programs.
- Misconceptions about wetlands as "wastelands" result in land conversion for development, agriculture, or drainage.

Conflicting Interests Among Stakeholders

- Governments, businesses, and communities often have competing priorities regarding wetland use.
- Economic interests (e.g., agriculture, urban development) frequently conflict with conservation goals.
- Lack of stakeholder dialogue leads to tensions and ineffective policies.

Limited Legal and Institutional Support

- Many national and local policies do not include clear frameworks for community participation in wetland governance.
- Weak enforcement of conservation laws allows illegal activities like pollution, encroachment, and overextraction to continue.
- Indigenous land rights are often not recognized, preventing traditional communities from managing wetlands sustainably.

Financial Constraints

- Many community-led conservation projects lack funding for long-term sustainability.
- Governments and donors often prioritize short-term infrastructure projects over ecosystem protection.
- Limited access to grants, subsidies, and investment programs hinders effective community involvement.

Strategies for Effective Community and Stakeholder Engagement

To overcome these challenges, inclusive and participatory governance models must be implemented, ensuring that wetlands are managed in a way that benefits both ecosystems and people.

Strengthening Community-Based Wetland Management

- Establish community-led conservation programs that empower local groups to take ownership of wetland protection.
- Provide financial incentives, training, and technical support for wetland restoration projects.
- Recognize indigenous land rights and involve traditional leaders in decision-making.

Enhancing Public Awareness and Education

- Implement school-based environmental education programs to foster long-term community engagement.
- Develop outreach campaigns and media initiatives to highlight wetland benefits and threats.
- Organize workshops, guided wetland tours, and citizen science projects to encourage active participation.

Promoting Collaborative Decision-Making

- Establish multi-stakeholder wetland committees that bring together government agencies, businesses, and communities.
- Encourage public participation in wetland policy-making through consultations and feedback mechanisms.
- Facilitate conflict resolution processes to balance conservation and development interests.

Expanding Financial and Market-Based Incentives

- Implement PES programs to compensate communities for wetland conservation.
- Encourage businesses to invest in wetland restoration through CSR initiatives.
- Develop ecotourism and sustainable livelihood projects that generate income while preserving wetlands.

Conclusion

Wetlands are among the most valuable and vulnerable ecosystems on Earth, providing essential services such as water filtration, flood control, carbon sequestration, and biodiversity conservation. As NbS, wetlands offer cost-effective and sustainable alternatives to engineered infrastructure, supporting climate resilience, water security, and sustainable development. However, despite their significance, wetlands continue to be degraded and lost at an alarming rate due to urban expansion, pollution, unsustainable land use, and climate change.

To ensure the long-term protection and restoration of wetlands, a comprehensive approach that integrates science, policy, economics, and community engagement is required. This book has explored the multiple roles wetlands play in climate resilience, water management, biodiversity conservation, sustainable development, and urban planning. It has also examined the economic value of wetlands, highlighting how they provide critical services at a lower cost than many engineered solutions.

Key Takeaways

- **Wetlands are crucial for climate adaptation and mitigation**: They act as natural buffers against floods, store carbon more efficiently than forests, and support ecosystems that regulate water cycles.
- **Biodiversity thrives in wetlands**: These ecosystems are hotspots for diverse plant and animal species, many of which are endangered or migratory.
- **Sustainable water management depends on wetlands**: They improve water quality, recharge groundwater, and balance competing water demands.
- **Wetlands support the SDGs**: Their protection aligns with global efforts to enhance food security, reduce poverty, and promote sustainable urbanization.

- **Restoration and conservation are economically beneficial**: Investing in wetland restoration reduces the need for costly flood defenses, water treatment plants, and climate adaptation infrastructure.

Challenges and Future Directions

Despite their benefits, wetland degradation persists due to conflicting land-use priorities, weak enforcement of environmental laws, and insufficient financial investment. Many wetlands lack legal protection, while those that are protected often face governance challenges, pollution, and encroachment. Strengthening policy frameworks, funding mechanisms, and cross-sector collaboration is essential to ensuring their preservation.

Moving forward, wetland conservation must be prioritized at all levels of governance—from international agreements like the Ramsar Convention to local urban planning initiatives. Governments must integrate wetlands into national climate policies, disaster risk reduction strategies, and water resource management plans. Meanwhile, the private sector can play a greater role by investing in wetland restoration projects, carbon offset programs, and sustainable land-use practices.

Additionally, community engagement and education will be critical in ensuring that wetland conservation efforts are inclusive and sustainable. Indigenous communities, farmers, and urban populations all depend on wetlands in different ways, making their participation vital for long-term stewardship.

Final Thoughts

Wetlands are not just passive landscapes; they are active systems that sustain life, regulate climate, and support economies. Protecting and restoring them is not just an environmental necessity but an economic and social imperative. By recognizing the true value of wetlands and integrating them into policy, finance, and urban planning, societies can ensure that these ecosystems continue to thrive, providing benefits for both people and the planet.

www.ingramcontent.com/pod-product-compliance
Lightning Source LLC
Chambersburg PA
CBHW071557200326
41519CB00021BB/6789